MICRO-ORGANISMS AND BIOTECHNOLOGY

Other titles in the Project

Biology Martin Rowland
Applied Genetics Geoff Hayward
Applied Ecology Geoff Hayward

Physics Robert Hutchings
Telecommunications John Allen
Medical Physics Martin Hollins
Energy Robert Hutchings and David Sang
Nuclear Physics David Sang

Design Electronics Bill Phillips

UNIVERSITY OF BATH • MACMILLAN SCIENCE 16–19 PROJECT

Project Director: J. J. Thompson, CBE

MICRO-ORGANISMS AND BIOTECHNOLOGY

JANE TAYLOR

First published 1990

Published by
MACMILLAN EDUCATION LTD
Houndmills, Basingstoke, Hampshire RG21 2XS
and London
Companies and representatives
throughout the world

Typeset by Typematter Graphics, Basingstoke, Hampshire

Printed in Hong Kong

British Library Cataloguing in Publication Data
Taylor, Jane
Micro-organisms and biotechnology
1. Industrial microbiology
I. Title II. Series
660′ .62
ISBN 0–333–48320–0

Contents

The Project: an introduction

The **University of Bath · Macmillan Science 16–19 Project,** grew out of a reappraisal of how far sixth form science had travelled during a period of unprecedented curriculum reform and an attempt to evaluate future development. Changes were occurring both within the constitution of 16–19 syllabuses themselves and as a result of external pressures from 16+ and below: syllabus redefinition (starting with the common cores), the introduction of AS-level and its academic recognition, the originally optimistic outcome to the Higginson enquiry; new emphasis on skills and processes, and the balance of continuous and final assessment at GCSE level.

This activity offered fertile ground for the School of Education at the University of Bath and Macmillan Education to join forces with a team of science teachers, drawn from a wide spectrum of educational experience, to create a flexible curriculum model and then develop resources to fit it. This group addressed the task of satisfying these requirements:

- the new syllabus and examination demands of A- and AS-level courses;
- the provision of materials suitable for both the core and options parts of syllabuses;
- the striking of an appropriate balance of opportunities for students to acquire knowledge and understanding, develop skills and concepts, and to appreciate the applications and implications of science;
- the encouragement of a degree of independent learning through highly interactive texts;
- the satisfaction of the needs of a wide ability range of students at this level.

Some of these objectives were easier to achieve than others. Relationships to still-evolving syllabuses demanded the most rigorous analysis and a sense of vision – and optimism – regarding their eventual destination. Original assumptions about AS-level, for example, as a distinct though complementary sibling to A-level, needed to be revised.

The Project, though, always regarded itself as more than a provider of materials, important as this is, and concerned itself equally with the process of provision – how material can best be written and shaped to meet the requirements of the educational market-place. This aim found expression in two principal forms: the idea of secondment at the University and the extensive trialling of early material in schools and colleges.

Most authors enjoyed a period of secondment from teaching, which allowed them not only to reflect and write more strategically (and, particularly so, in a supportive academic environment) but, equally, to engage with each other in wrest-ling with the issues in question.

The Project saw in the trialling a crucial test for the acceptance of its ideas and their execution. Over one hundred institutions and one thousand students participated, and responses were invited from teachers and pupils alike. The reactions generally confirmed the soundness of the model and allowed for more scrupulous textual housekeeping, as details of confusion, ambiguity or plain misunderstanding were revised and reordered.

The test of all teaching must be in the quality of the learning, and the proof of these resources will be in the understanding and ease of accessibility which they generate. The Project, ultimately, is both a collection of materials and a message of faith in the science curriculum of the future.

J.J. Thompson
January 1990

How to use this book

The aim of this book is to describe the nature, importance and applications of micro-organisms and cell technology. This is an area of biology which has developed very fast in recent years and which is likely to have an increasing impact on everyday life. Its growing importance is reflected in the greater emphasis given to these topics in current syllabuses, for although micro-organisms and their role in health and disease has been an optional study for many years, new A- and AS-level syllabuses also include biotechnology and cell technology products.

Micro-organisms and Biotechnology is written for students who are following an Advanced level or AS-level course in biology, particularly those following an option in micro-organisms and biotechnology. It will also be helpful to those following BTEC courses in health studies, home economics and some chemistry courses.

There are three themes: the first deals with the nature and growth of isolated cells and micro-organisms; the second with environmental and industrial microbiology, and the final theme covers micro-organisms and disease in plants and animals. Although it may be easier to start at the beginning of the book and work your way through, you do not have to do this. You can use just a couple of chapters or a single theme without having to study all other parts, but if you do so, you should read some sections of chapter one first. These sections of chapter one are listed on the pages introducing each theme, under the heading pre-requisites. The chapters may all contain a little more information than you are likely to have to learn, but even if you do not need to study a particular section in depth you will find that reading through it helps to broaden your understanding. You should consult your syllabus to see which topics need to be covered in depth and you will find the index and contents list help locate particular items.

The material is arranged to assist you in organising your studies. On the page opposite you will find information about the different features which will help you get the best out of the book.

In examinations you are expected to be able to select and use correct scientific terminology, and may by penalised if you do not do so; the material is written to help you understand and use technical terms. When an important word or technical term is used for the first time it is printed in heavy type, **like this**. It may be defined, or its meaning explained in the surrounding two or three sentences. It is not explained again if it is used later in the book. If you are just using one or two chapters of the text and come across a technical term you don't understand you can use the index as a glossary; look up the term in the index and the number in heavy type refers to the page with the definition on.

I would like to thank the many students and teachers who used trial material and sent me their invaluable comments and suggestions. In particular, I would like to thank my own A-level students who have patiently endured being guinea pigs for all kinds of exercises and ideas. I would also like to thank Joan Angelbeck for her help, advice and limitless patience in putting this book together; Mary Waltham for getting it all started and John Wormald and Mrs Jenny Jones for their help and encouragement while writing this book. When you use this text I hope that you will not only learn a lot but also get some enjoyment out of reading about the world of micro-organisms and how we exploit them.

Learning objectives

Each chapter starts with a list of learning objectives which outline what you should gain from the chapter. They are statements of attainment and often link closely to statements in a course syllabus. Learning objectives can help you make notes for revision, especially if used in conjunction with the summaries at the end of the chapter, as well as for checking your progress.

Questions

In-text questions occur at the end of sections of work. Some test recall and need only a word or phrase in answer; others need a longer explanation. These questions are to help you memorise the facts you need. Others ask you to use knowledge to interpret data and some are open-ended with many possible answers. The questions start very simply and become more demanding, and although you can either do these questions as you go along while the information is fresh in your mind, or save them to do at the end of a bigger section, the longer you leave them the harder they may be to answer. Answers, or answer outlines, are provided for guidance where appropriate.

Case studies

These are comprehensions, on topics related to work in the main text, to test your understanding of scientific writing.

Practical work

Although some practical techniques are illustrated there are no specific practical exercises. However, many of you will have to carry out an independent investigation so there are some ideas for independent work at the end of the book. Some of these do not require bench practical work.

Boxes

In each chapter some information has been placed in a box. This is extension work which you do not have to learn, but which has been included to help your understanding of microbiology and its applications.

Margin notes

These short notes summarise important points which you must remember.

Summaries

Each chapter ends with a brief summary of content. These summaries, together with the learning objectives, should give you a clear overview of the subject and allow you to check own progress.

Further reading

A list of books is included at the end of the book for those who wish to widen their understanding.

Acknowledgements

The author and publishers wish to thank the following for permission to use copyright material:

Joint Matriculation Board for exam questions from past examination papers.

The author and publishers wish to acknowledge, with thanks, the following photographic sources:

Agricultural Development and Advisory Service, MAFF *pp 149, 152;* Agricultural Genetics Company Limited *p 98 upper;* Heather Angel, Biofotos *p 46 lower;* Applikon Instruments *p 38;* Biophoto Associates *pp 7 left and right, 11 right, 19 upper and lower, 21, 23, 27 upper, 32, 47 middle, 120, 121 lower, 123, 171;* Anthony Blake *pp 55, 66 top;* Reproduced by courtesy of the Centre National D' Etudes Sur La Rage Et La Pathologie Des Animaux Savages *p 172 middle and lower;* Bruce Coleman *pp 34, 151 upper;* P Dunnill *p 102;* Mary Evans Picture Library *pp 56 lower, 128;* ExacTech *p 105 left;* Farmers' Weekly Picture Library *pp 86, 117, 140, 145 top, 165 right;* Ford Motor Company *p 87;* Geoscience Features Picture Library *pp 109, 150 upper, 151 lower, 153;* Philip Gordon/Reflex *p 59;* Philip Harris Biological Ltd *pp 83 upper, 132;* Holt Studios Ltd *pp 146 upper and lower, 150 lower, 158 left and right;* MAFF *p 154 lower;* National Coal Board *p 83;* National Dairy Council *p 65;* Oxford Scientific Films *p 133;* Panos Pictures *p 66 bottom;* Charles River Biotechnical Services *p 91;* Ann Rohan *p 3 upper and lower;* Science Photo Library *pp 1, 5 left, centre and right, 11 left and centre, 14 upper and lower, 16, 17, 19 centre, 25, 28, 29, 47 left and right, 51, 70 left and right, 88 lower, 92, 125, 131, 145 lower, 165 left, 172 top, 174;* South American Pictures *pp 89, 90;* Stilton Cheese Makers Association *p 56 upper;* Unipath Ltd *p 113;* C James Webb *pp 13, 20 upper and lower, 40, 44, 52, 53, 61, 83 lower, 99 left, 104, 105 upper, 106, 124, 134, 138, 157, 164, 169;* The Wellcome Foundation *p 98 lower;* Wildlife Matters *pp 80, 88 upper;* World Health Organisation *pp 99 right, 137, 163, 167, 177;* Zefa Picture Library *p 58.*

Every effort has been made to trace all the copyright holders but if any have been inadvertently overlooked the publishers will be pleased to make the necessary arrangement at the first opportunity.

Theme 1

THE NATURE OF MICRO-ORGANISMS

Micro-organisms are those organisms which are too small to be seen with the naked eye. They are mainly single-celled but that is the only feature that they have in common, even then most fungi are multicellular. The invisible world of micro-organisms has been explored since the invention of the microscope in the seventeenth century but real understanding has only developed in the last hundred years and new information is arriving every day.

A major development is the systematic commercial exploitation of micro-organisms. Though we have long used yeasts and lactic acid bacteria in food manufacture, and exploited the nitrogen-fixing capacities of bacteria in agriculture, the last few decades have seen an expansion in our understanding of how micro-organisms and isolated cells can be grown and how their metabolism is controlled.

Prerequisites

An understanding of GCSE chemistry; it is advantageous to have studied A-level cell biology and respiration.

Chapter 1

MICRO-ORGANISMS

LEARNING OBJECTIVES

After studying this chapter you should be able to:

1. appreciate the main steps in the history of the study of micro-organisms;
2. understand the units used to describe micro-organisms;
3. appreciate the nature of different groups of micro-organisms.
4. describe the principal features of the main groups of micro-organisms.

1.1 WHY STUDY MICRO-ORGANISMS AND BIOTECHNOLOGY?

Micro-organisms have an enormous impact on our lives. They can bring about changes in the environment and cause disease in people, animals and plants. The scientific investigation of micro-organisms has provided us with an understanding of how they work and of their abilities, and we are now able to exploit these abilities to perform all kinds of chemical activities. In a similar way to micro-organisms, isolated animal and plant cells can be grown, and many different uses have been found for these too.

Biotechnology is a rapidly expanding area which uses micro-organisms, cells and cell products for industrial and commercial use. New techniques have been developed which have led to advances in many areas of medicine, agriculture and commerce. Today everybody's life is touched in some way by microbial or cell technology. The end of the 1980s has seen, for example, the introduction of protein made by micro-organisms into 'meat' pies, cars driven by fuel made by microbial fermentation of agricultural waste, and millions of people using microbial fermenters to provide power. If you travel to other countries you may be given genetically engineered antigen-based vaccines, and you probably use bacterial enzymes to get your washing clean. Monoclonal antibodies are used in many different diagnostic tests, from 'do-it-yourself' pregnancy tests to checking what kind of meat is in a hamburger.

Each chapter in this book deals with a different aspect of the growth and activity of isolated cells and micro-organisms.

1.2 THE DEVELOPMENT OF BIOTECHNOLOGY

We know that micro-organisms were used in brewing and baking in the Middle East many thousands of years ago, but their activity was not recognised at the time because micro-organisms are too small to be seen with an unaided eye. It needed the development of the microscope in the seventeenth century before the nature of very small organisms could be investigated. During the next two hundred years investigators began to explore the world of micro-organisms and cells, drawing and describing what they saw. No-one, however, knew where micro-organisms came from and there were many arguments about the origin and nature of these microscopic organisms.

(a)

(b)

Fig 1.1 The two great figures of 19th century microbiology. Louis Pasteur **(a)** and Robert Koch **(b)**, shown here on commemorative medals, worked in great rivalry.

The nineteenth century was a time of great advances in scientific knowledge and experimentation, including the investigation of micro-organisms. Diseases of plants were investigated and preventative treatments developed. At the beginning of the century the life expectancy of people in Western Europe was between 30 and 40 years, and by the end of the century it was extended by decades. This increased longevity came largely from the recognition of the need to provide more hygienic living conditions, adequate nutrition and good water supplies and also from a better understanding of the role of micro-organisms in disease. Most of the credit goes to two men, shown in Fig 1.1, and the scientists they gathered around them. Louis Pasteur was French; Robert Koch, slightly younger, was German. Their governments had been at war when they were both young men and they met only once. Joseph Lister, the British surgeon who introduced the use of antiseptics, arranged the meeting, but it was reported as being a dismal failure and they worked in great rivalry, and with some animosity, for the whole of their lives. However, between them they can be credited with the discovery of many causes of disease; how micro-organisms cause illness; how to grow, examine and investigate micro-organisms; the systematic development of vaccines and anti-toxins; and the development of hygienic practices to prevent the spread of infectious diseases.

By the start of the twentieth century the role of micro-organisms in the environment became clearer and new micro-organisms such as viruses were discovered. Work was done on therapeutic agents and antibiotics, and synthetic drugs were developed to treat sick people. The practice of routine vaccination together with clean water supplies reduced the incidence of many killer diseases in Europe. The tanning industry refined the use of enzymes in leather preparation and the first 'biological' washing powder was formulated, though it was not very successful. In the middle of the century investigations into the biology of cells lead to the development of techniques for growing isolated animal and plant cells.

In the 1950s and '60s investigators learnt how cells are controlled, how inherited information is carried and how bacteria carry out genetic exchange without sexual reproduction. This allowed researchers of the 1970s to learn how to manipulate genetic material in cells and how to transfer information from one cell to another. Plant biologists could manipulate plant cells in different ways to generate large numbers of genetically identical plants and to carry out more effective breeding programmes for agriculture and horticulture. The 1970s and '80s led to a huge growth in the exploitation of cells, micro-organisms and their enzymes to carry out a variety of industrial production processes more cheaply and more effectively than ever before.

Table 1.1 shows the major steps in these studies. It is now possible to manipulate genetic instructions in cells to make them do things they don't normally do, and even to transfer genetic instructions from one kind of cell to another with relative ease. Our problems today are not so much in manipulating genetic material, but in persuading cells and organisms to grow in giant fermentation tanks.

Table 1.1. A time line of landmarks in microbiology.

17th C.	**Anton van Leeuwenhoek,** 1632-1723, discovers microscopic organisms in pond water using his newly developed microscope.
1683	Van Leeuwenhoek draws mouth bacteria.
1718	Lady Montague has her son inoculated against smallpox after observations in the Middle East.
	Abbè Lazaro Spallanzani, 1729-99, investigates the appearance of decay organisms in foodstuffs.
1796	**Edward Jenner** experiments with a vaccine of cowpox to prevent smallpox.
1802	Forsyth, gardener to George IV, uses lime sulphur to treat mildew on fruit trees.
1846	Rev. Berkeley states that potato blight in the Irish famine was caused by a fungus.

Table 1.1 cont.

1856	**Louis Pasteur,** 1822-1895, disagrees with Liebig and establishes that yeast is needed for fermentation, souring being caused by other material.
1863	Pasteur investigates wine souring and discovers that it is caused by an organism found in vinegar barrels. Develops pasteurisation.
	Anton de Bary shows potato blight to be caused by a fungus.
1864	Pasteur establishes that decay organisms are found as small organised 'corpuscles' or 'germs' in the air.
1865	Pasteur investigates silkworm disease and establishes that diseases can be transmitted from one animal to another.
	Joseph Lister reads Pasteur's work and starts using disinfectants in wound care and surgery.
1866	Pasteur develops a germ theory of disease.
1868	Davaine uses heat treatment to cure a plant of bacterial infection.
1873-6	**Robert Koch** investigates anthrax and develops techniques to view, grow and stain organisms and photographs them, aided by Gram, Cohn and Weigart.
1879	Pasteur grows weakened strains of organisms that cannot cause disease but protect against severe forms.
1880	Koch establishes a protocol to show that an organism causes a disease.
1882	Pasteur uses Koch's work to produce a vaccine against anthrax.
	Koch identifies the TB organism and hence the first human microbial disease
	Ilya Metchnikoff observes phagocytes surrounding micro-organisms in starfish larvae. Later develops a cell theory to explain the action of vaccines.
	Millardet develops Bordeaux mixture to treat powdery mildew on imported vines.
1884	Pasteur develops rabies vaccine.
1885	Human trials of rabies vaccine.
1885-95	Koch, **Petri,** Löffler, Yersin and **Erlich** identify many human disease-causing organisms.
	Behring develops diphtheria anti-toxins.
1886	J C Arthur demonstrates that pear fire blight is a bacterial disease.
1887	Pasteur gives up work through ill-health. Pasteur Institute opened.
1890s	Roux develops diphtheria vaccine.
1892	**Iwanowski** shows tobacco mosaic disease is a transmissible disease caused by an agent too small to be filtered out; the first plant virus disease is identified.
1895	**Winogradski** demonstrates nitrogen fixation in the absence of oxygen by *Clostridia* bacteria.
1898	Löffler and Frosch identify the first animal viral disease, foot-and-mouth.
1900	**Walter Reed** identifies yellow fever as the first human viral disease.
1901	**Beijerinck** identifies free-living aerobic nitrogen fixers.
1906	Paul Erlich works on his 'magic bullet'. Investigates atoxyl compounds, finds Salvarsan – the first chemotherapeutic agent.
1908	Calmette and Guerin develop a vaccine against TB. BCG used in 1921.
1915	**Twort** describes bacteriophages: viruses that attack bacteria.
1928	**Alexander Fleming** observes bacteria destroyed around an area of *Penicillium* mould growth.
1930s	Gerhard Domagk investigates Prontosil, a dye. Later the Trefouels derive a sulphanilamide drug.
	Isolated plant tissues cultured by a variety of workers. Organo-sulphur fungicides developed.
1935	**Stanley** crystallises tobacco mosaic virus.
1938	**Florey** and **Chain** isolate the anti-microbial agent penicillin.
1939	**Gauteret** grows carrot callus cultures.
1944	Waksman isolates streptomycin; effective against TB.
1940s	Isolated animal cell cultures grown in laboratories.
1950s	**Earle** and Enders grow monkey, mouse and chick cells in cell cultures. Gey develops the HeLa human cell line.
1959	**Reinart** regenerates plantlets from carrot callus culture.
1961	**Jacob and Monod** establish the basis of genetic regulation.
	Systemic fungicides developed.
1966-9	Restriction enzymes that snip DNA investigated.
1960-70s	Plant cell protoplasts cultured.
	Viroids identified.
1965	Harris and Watkins fuse mouse and human cells together.
1975	**Kohler and Milstein** fuse cells together to produce monoclonal antibody.
mid-1970s	Tonegawa clones mouse DNA.

1.3 WHAT ARE MICRO-ORGANISMS?

The group of living things known as 'micro-organisms' is an arbitrary one, defined as organisms which are too small to be seen unaided. This is not always true, though, for example most people have eaten mushrooms, which are fungal fruiting bodies. As well as fungi, the group includes bacteria, viruses, protozoa and green algae. In this book the emphasis is on those micro-organisms which are important for industrial, commercial or health reasons.

Micro-organisms have a wide range of characteristics. They are able to colonise almost all ecological niches including such hostile places as the Arctic, inside engine fuel lines, coal mines and hot springs. Most of them are **unicellular**; that is, they have only one cell which has to carry out all the functions necessary for life: feeding, respiring, reproducing, excreting metabolites and so on. Some are motile and have structures which allow them to move about independently. Some micro-organisms, for example the fungi, are **multicellular**. These may have **differentiated** structures, that is different cells are specialised for particular tasks such as reproduction. Others may have **colonial** growth, where individual cells may grow on their own, but usually in close association with others, and they gain benefit from the association. The dividing lines between unicellular, colonial and multicellular structures are far from clear.

1.4 SEEING MICRO-ORGANISMS

The smallest objects that can be seen without a microscope are larger than 0.1 mm, though at this size little detail can be discriminated. An optical microscope with good lenses gives a magnification of up to 2000 times, which allows us to see some of the external features of bacteria and protozoa. The use of stains taken up by particular chemical components and phase contrast microscopy have added to our knowledge of cell structure. However, there is a limit to what can be seen with an optical microscope. Cell walls, the presence of a bounding cell membrane and structures such as cilia and flagella can be seen, as can nuclei, very large chloroplasts and stained mitochondria. Little detail of their structure can be distinguished however – even greater magnification is needed for this.

Bacteria can be seen with an optical microscope. The electron microscope reveals details of cell structure.

An electron microscope is used to see details of ultrastructure. Various techniques can be used to clarify certain structural details and specimens can be treated in different ways. For example, if a specimen is frozen and then fractured it often splits along lines of crystallisation or along membranes thus revealing the nature of inner structural planes. A scanning electron microscope uses another technique to form an image which reveals something of the three-dimensional nature of a cell or micro-organism. A comparison of the images from each technique can be seen in Fig 1.2.

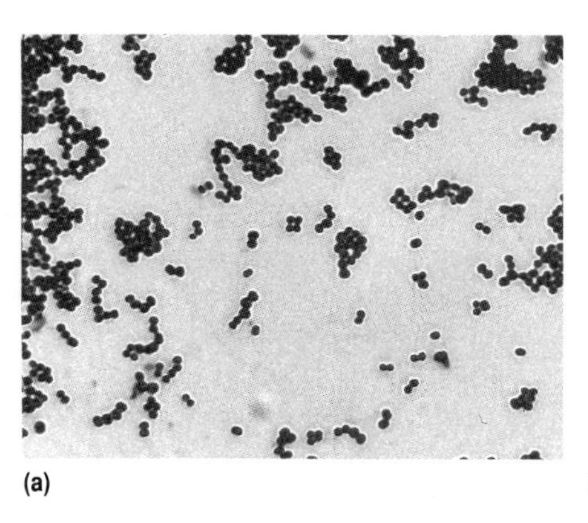
(a)

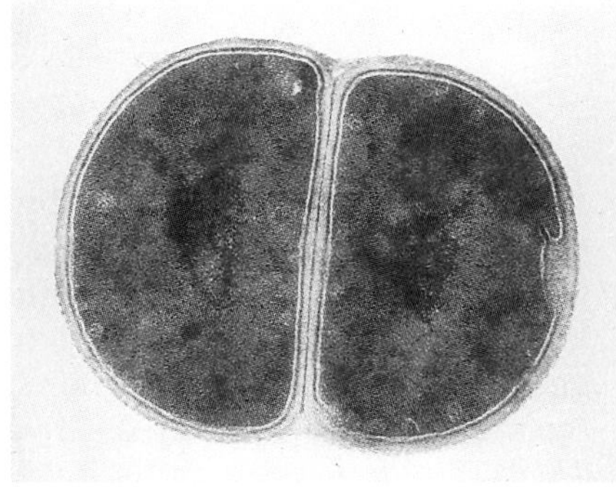
(b)

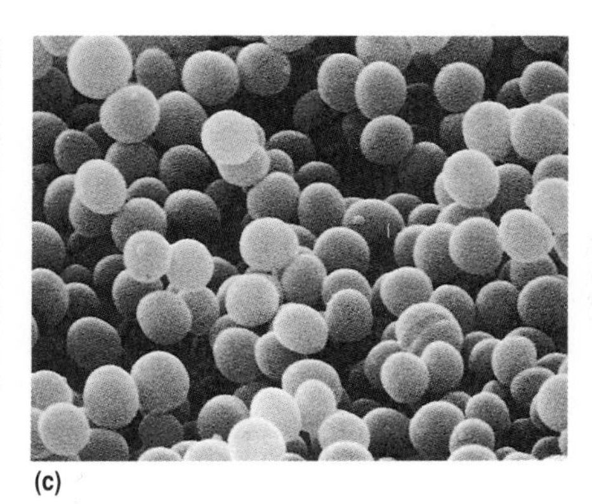
(c)

Fig 1.2 The same organism viewed by three different techniques; **(a)** is a smear of *Staphylococcus pyogenes aureus*, stained and photographed using a light microscope, originally × 500. Very little detail can be seen. An electron microscope was used to take **(b)**, a newly divided *pyogenes aureus* cell, with an original magnification of x 30 000. The wall and some internal detail can be seen. A scanning electron microscope was used for **(c)** at a similar magnification to (b). The shape and surface detail of the bacteria can be seen.

Units

The units used to measure mice, sepals or the size of an elephant's ear are inappropriate for micro-organisms. The unit used most often is the micrometre (μm). One micrometre is one thousandth of a millimetre, or one millionth of a metre (10^{-6} m). Most animal and plant cells, protozoa and algae are between 0.5 and 20 μm in diameter and can be seen with an optical microscope. However the largest bacteria are less than 4 μm long and viruses are at least ten times smaller than this. For these very small micro-organisms the nanometre (**nm**) is used. This is one millionth of a millimetre (10^{-9} m). Fig 1.3 illustrates the sizes of various cells. An optical microscope will allow us to see things as small as 0.2 μm. An electron microscope, with a resolving power down to 0.3 nm, will allow us to see objects a hundred times smaller than an optical microscope.

Micro-organisms are measured in micrometres (μm).

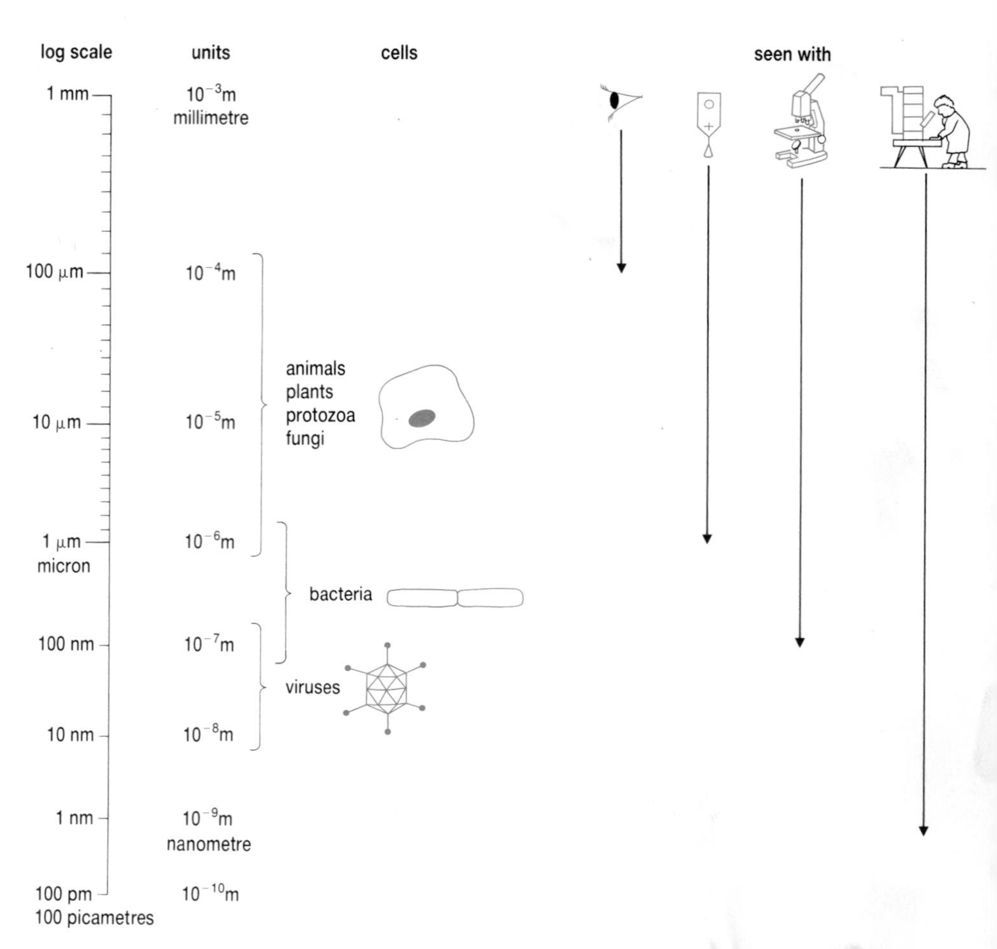

Fig 1.3 A comparison of the sizes of micro-organisms and cells.

1.5 PROKARYOTES AND EUKARYOTES

Micro-organisms are grouped by common features of their structures and components. There are still gaps in our knowledge of some organisms, so they are grouped with those with which they have most similarities. As we learn more about an organism, for example more about its metabolism or genetics, we can assign it more accurately. This may sometimes involve renaming it, for example *Ferrobacillus ferrooxidans* is now known as *Thiobacillus ferrooxidans* after it was moved to the genus *Thiobacillus*.

Prokaryotes and eukaryotes

Micro-organisms are divided into two main groups based on their internal structure, which is much more complex in some than in others. The key feature is the presence or absence of structures made of membranes. The modern descendants of the earliest living organisms do not have membranous structures such as chloroplasts or mitochondria. These are the **prokaryotes**, which include bacteria and blue-green bacteria.

There are two types of cells: prokaryotes without internal membrane systems and eukaryotes with complex internal structures.

After internal membranes evolved, cellular activities could be compartmentalised and more metabolic activities could take place at the same time without interfering with each other. Organisms with internal membranes in their cells are **eukaryotes**. Eukaryotic cells contain characteristic organelles and are between 1 and 100 μm in diameter, which is ten times larger than a typical prokaryotic cell. Fig 1.4 (a) and (b) are electron micrographs of a prokaryotic and eukaryotic cell; the lack of membrane structures can be seen easily. Eukaryotes include protozoa, fungi, algae and all more complex organisms.

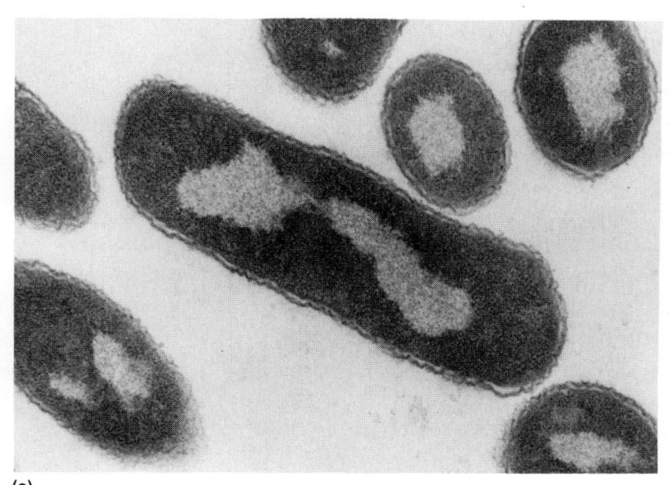

(a)

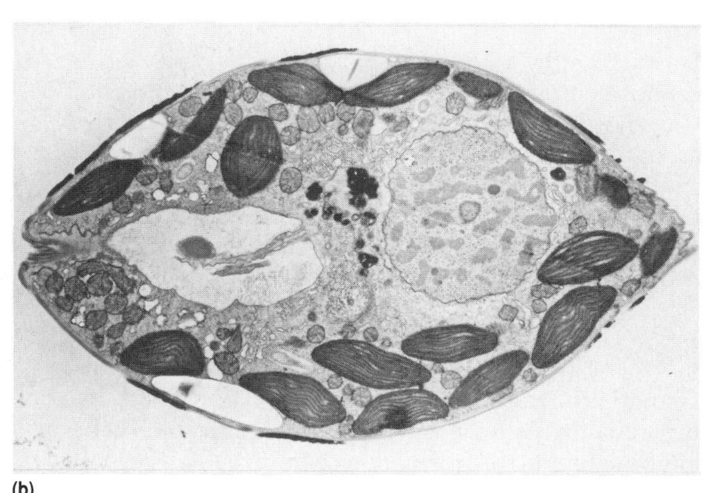

(b)

Fig 1.4 Eukaryotic cells have different internal structures to prokaryotes. *Escherichia coli* **(a)** is prokaryotic, a cell wall can be distinguished and a paler zone of nucleoplasm. In contrast *Euglena gracilis,* **(b)** is eukaryotic with chloroplasts, a nucleus, mitochondria and an endoplasmic reticulum, which can be seen in the electron micrograph.

Table 1.2 Comparison of prokaryotic and eukaryotic features.

Eukaryotic cells	Prokaryotic cells
Surface membrane	
A phospholipid bi-layer with proteins and glycoproteins in a fluid mosaic. Controls the entry and exit of molecules which enter or leave by diffusion, or are assisted by the cell membrane using energy from respiration. Large molecules can only pass by endo- and exocytosis. Surface area may be enlarged by projections such as microvilli.	Has a similar structure; the only membrane present. It may be folded inwards enlarging the surface available for materials to exchange. Functions such as respiration and photosynthesis are located on it. Passage of materials is controlled by the membrane but unusually large molecules such as fragments of DNA and enzymes pass through. Endocytosis is not observed.
Cell wall	
Cell walls, if present, are made of cellulose except in fungi which have cell walls made of a number of substances including chitin.	Most have a cell wall. Acts as a barrier and strengthener, resisting water entry by osmosis. Thick, but not always easy to tell where the cell wall ends and the membrane begins. Murein is a major component.
Internal membrane structures	
Cytoplasm is subdivided by a network of membranes called the endoplasmic reticulum, continuous with the surface membrane and with membranes round the nucleus. Passage of materials across these is similar to movement across the surface membrane. Internal membranes may be studded with **ribosomes**.	No internal membranes; ribosomes are smaller and scattered, sometimes clustered, in the cytoplasm.

Table 1.2 cont.

Membrane-bounded bodies Several types, each specialised for particular functions. Most of the enzymes associated with energy production are in **mitochondria. Chloroplasts** house photosynthesis. Chlorophyll and other pigments are found in the layers of thylakoids within them; the enzymes for carbon dioxide fixation and ATP production are between the layers. Secreting cells have **Golgi apparatus** which is involved in the packaging of materials for transport elsewhere or for release from the cell.	No membrane-bounded bodies. Some photosynthetic bacteria and blue-green bacteria may have simple sacs close to surface membranes housing photosynthetic pigments.
Nucleus Surrounded by a double membrane with pores. It contains chromosomes which carry the cell genome. Chromosomes are long molecules of DNA packaged around histone proteins. The nucleus is the site of RNA synthesis and editing. Some organelles carry some DNA, such as mitochondria and chloroplasts. The genome is duplicated before mitosis. Genetic exchange occurs during sexual reproduction.	There is a region of nucleoplasm which is the site of the genome but not separated. The genome is a loop of double-stranded DNA but no histones. It carries the genes needed for growth and development. Many bacteria have separate loops of DNA called **plasmids** carrying non-essential genes. The duplication of the genome and cell division may not be closely linked. Genetic exchange is not linked with reproduction but involves one organism taking up genetic material donated by another.
Osmotic control Protozoa regulate the water potential of the internal cell environment by the formation of intracellular vacuoles. These contain surplus water which is expelled to the external environment. Multicellular organisms use different methods.	Osmotic vacuoles are not made by prokaryotes.
Reproduction There are asexual methods which produce new cells without genetic exchange. Many organisms can reproduce sexually with the exchange of genetic material and this may be the only method available to some. The most common pattern is the production of specialised cells containing half the genome, **gametes**, by meiosis. When two gametes fuse the genome is completed. Many eukaryotic micro-organisms are haploid for most of their life cycle. In these, fusion of the gametes produced by mitosis results in a diploid zygote. Meiosis then occurs making haploid cells again.	Reproduction is usually asexual, taking place by a process of binary fission or budding.
Movement Many eukaryotic micro-organisms are motile: some move by the internal movement of cytoplasm, others have **flagella** or **cilia** to propel them through moist environments. Cytoplasmic streaming has been observed in some micro-organisms.	Prokaryotes do not move by the internal movement of cytoplasm. The cell wall would make this difficult anyway, but some organisms have a gliding motion. Some organisms have flagella, enabling them to swim through films of moisture.
Flagella Some cells have flagella rooted in the cytoplasm but projecting beyond the cell membrane and wall. Flagella have a pair of protein fibres extending down the length surrounded by nine pairs of fibres enclosed by membrane. The fibres contract to produce motion.	Flagella have two or three protein molecules entwined to produce a fibre rooted under the cell wall. Acts like a helical rotor to propel the organism through moisture films.

Though the main difference is in the presence or absence of internal membranes, there are other important differences between the two groups. These are shown in Table 1.2. Viruses cannot be included in either group as they do not have a cellular structure. Though some viruses are enveloped with an outer coating of cell membrane, the membrane is not their own: it is acquired from a host as the virus leaves the host cell.

QUESTIONS

1.1 List as many ways as you can in which micro-organisms affect you in your daily life. Revise the list after you have read more of the book.

1.2 Write down the following units in decreasing order of size: nanometre, kilometre, centimetre, micrometre, millimetre.

1.3 If you used a light microscope to examine a eukaryotic cell and compared it with a prokaryotic cell, what features would you be able to **see** that would distinguish the two types of cell?

1.6 THE MAIN GROUPS OF ORGANISMS

We are mainly concerned with the activity of bacteria, fungi and viruses but other groups are important too in any study of microbial life, so protozoa and unicellular green algae are included. The next section contains a very brief outline of the main groups and their importance; a more detailed study of each group can be found in later sections. Fig 1.5 illustrates the relationships of some of the micro-organisms discussed in this book.

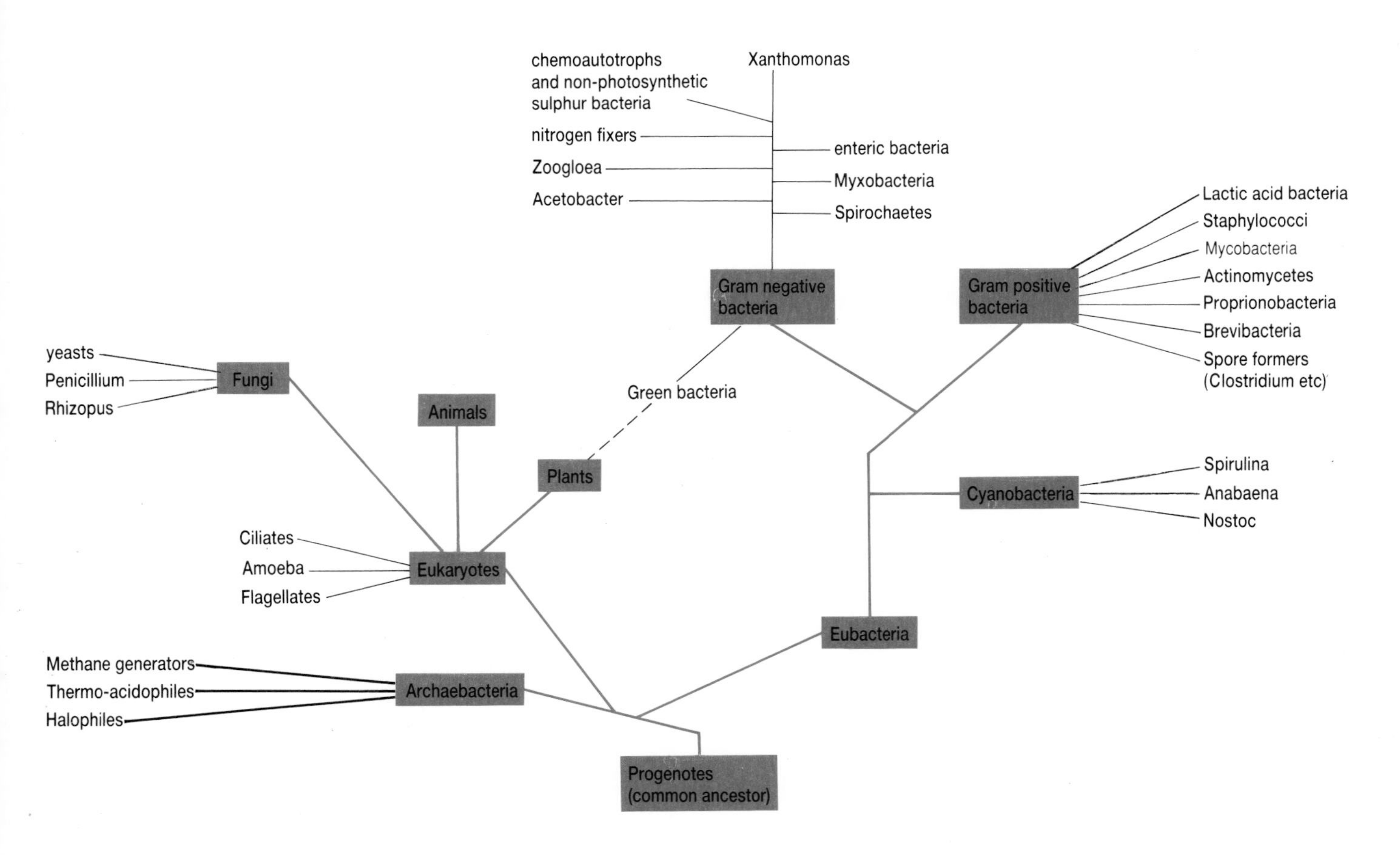

Fig 1.5 The relationships between micro-organisms.

Prokaryota

The Kingdom of **Prokaryota** is divided into two classes, Bacteria and Cyanobacteria, both prokaryote cell types. Bacteria are found in every ecosystem and grow in a wide range of environmental conditions. Many are saprophytic, degrading organic material in soil and water or as parasites of animals and plants; but there are also photosynthetic bacteria and bacteria which gain their energy from inorganic chemical oxidations. In contrast the cyanobacteria are all photosynthesisers. The photosynthesisers in both groups are among the main primary producers in areas where vascular plants do not grow.

Most bacterial species are single-celled, surrounded by a strong cell wall which often has a protective coating. They vary in size enormously, the largest species are thousands of times bigger than the smallest. They can carry out a wide range of biochemical activities, many of which are unique to bacteria. In a favourable environment bacteria proliferate asexually and quickly colonise any available substrate. They do not reproduce sexually, but can exchange genetic material in a variety of ways. Some can move independently, but many are reliant on wind or other passive mechanisms for dispersal.

Many cyanobacteria, though like bacteria in most features, are filamentous. However each cell is able to lead an independent existence and reproduce if the chain of cells becomes fragmented. They photosynthesise using a range of pigments arranged in membranes, these pigments being responsible for their range of colours. Some species can fix nitrogen in nitrogen-poor environments. This involves the specialisation of cells in the chain which, as a result of this, lose their ability to support and reproduce themselves.

Protoctista

The Kingdom of **Protoctista** includes the Protozoa, the Euglenoids and the Chlorophyta, or Green Algae. Each group has its own characteristics, but some micro-organisms have features belonging to more than one group. The protoctists are eukaryotic and unicellular, though some algae live as colonial forms.

Most protozoa are big enough to see with a good optical microscope, up to 5 μm across. It is a diverse group with many interesting features, but they do have some characteristics in common, for example protozoa have a tough cell membrane, but no cell wall. They are heterotrophic, often with specialised feeding mechanisms, and most can move independently, coordinating different parts of the cell in search of food or responding to stimuli. Protozoa reproduce sexually as well as asexually and can reach large numbers in favourable surroundings.

Euglenoids have some protozoan and some algal features. They are single-celled with a tough cell membrane but have no cell wall. Most are photosynthesisers and they can swim around using a flagellum. In sunlight they photosynthesise using chloroplasts, but in the dark they become heterotrophic and absorb soluble organic nutrients from the surrounding medium. There are some colourless euglenoids which cannot photosynthesise and these are entirely heterotrophic.

The unicellular green algae are more plant-like than the Euglenoids as most have a cellulose cell wall and are completely photosynthetic, but many are also flagellate. Some members of the group are colonial, filamentous or multicellular. Not all unicellular algae have cellulose walls – desmids and diatoms have silicate shells. Fig 1.6 shows some of the intricate shell shapes found in this group.

(a)

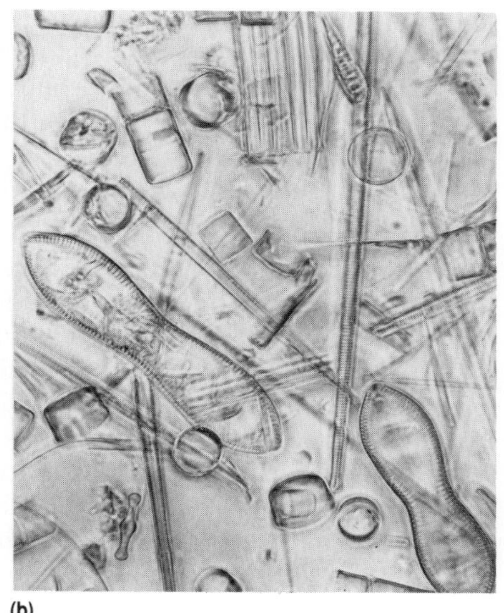

(b)

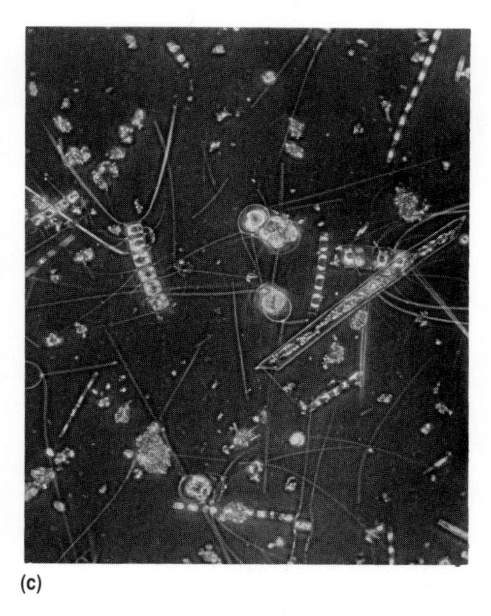

(c)

Fig 1.6 (**a**) Two sorts of desmid with quite different shapes. (**b**) Diatoms also come in many shapes and sizes, each has a box-like shell around it. Plankton (**c**) is a mixture of diatoms and desmids with animal larvae, cyanobacteria and alga.

The Fungi

Fungi are a diverse group of heterotrophs which obtain nutrients by secreting enzymes into their immediate environment and then absorbing the soluble products through the cell wall. The least complex fungi are protozoa-like, but most are multicellular. A fungal 'body' is made of filaments called **hyphae** which form a widespread thread-like mass throughout the substrate it is growing on; parts may be seen as a fluffy mass. Special hyphae are used to produce large numbers of sexual and asexual spores which disperse widely.

Fungi are widespread in soil and water. They are aerobic and very tolerant of acidity and dryness and this allows them to grow in substances with very low water potentials such as jam, stored grain and wood. Most fungi live on dead organic material and many produce a very wide range of enzymes that includes cellulases and lignases. They play a key role in the cycling of organic matter. Some fungi are specialised as parasites and are important plant pathogens, but there are some which parasitise animals.

Viruses

Viruses are very different from any of the previous groups as they do not have a cellular structure. Outside cells they are inert particles called **virions** made of a protein coat wrapped round a nucleic acid molecule. Virions can infect susceptible cells but have no other abilities, and are entirely dependent on passive dispersal, or a vector, to transmit them from one host to the next. Only when they gain entry to a living cell can they show properties of living organisms. Inside cells they are very small obligate intracellular parasites which can usurp normal cell controls and make cells divert their efforts into duplicating viral components. There are viral parasites of all groups of living organisms. Even bacteria are infected by viruses, such viruses are called **bacteriophages**, or phage for short.

Naming micro-organisms

The first part of a micro-organism's name is its genus, (written with an initial capital letter) and is shared with close relatives. The second part, or species, name is unique to a species and is based on features such as its metabolism, source, the discoverer, or any disease it causes. Some species have a third part to the name which refers to a particular strain. The second and third parts of the name do not have a capital letter. Viruses are not named in the same way, usually they are named after the disease they cause and the organisms affected.

1.7 BACTERIA

Bacteria are single-celled and range from 0.5 µm to 5 µm long; one exceptional group, the Actinomycetes, form filaments. Bacteria were first grouped by shape and movement, and subdivided by other characteristics such as growth pattern or the stains taken up. Fig 1.7 illustrates these original groups: the rod-shaped **bacilli**, for example *Bacillus pasteurii*: the round **cocci**, for example *Streptococcus lactis*; and the **spiral bacteria** such as *Spirillum* and *Spirochaeta*. It soon became obvious that these three groups were inadequate and now there are two main groups, which are each subdivided into smaller ones by metabolic, structural and genetic features. The **Archaebacteria** are thought to be very old – possibly the earliest life forms, and include the methane generating bacteria, the halophiles or salt-tolerant bacteria, and the thermoacidophiles which are adapted for life in hot acid places. The **Eubacteria** developed later and include, amongst others, photosynthetic bacteria, sulphur bacteria, slime bacteria, pseudomonads, the enteric bacteria, all Gram positive bacteria and the actinomycetes.

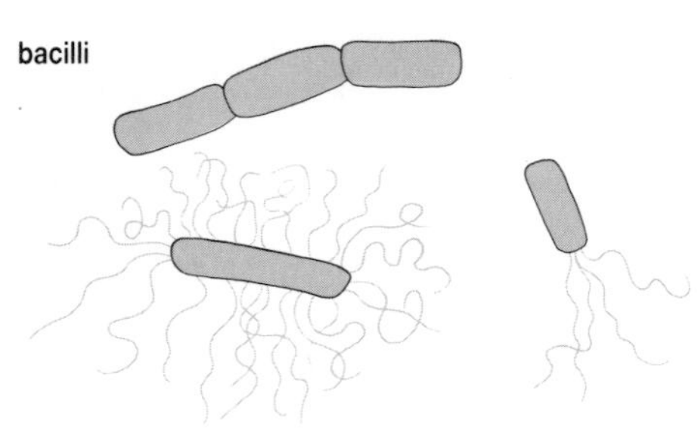

single, chains with and without flagella

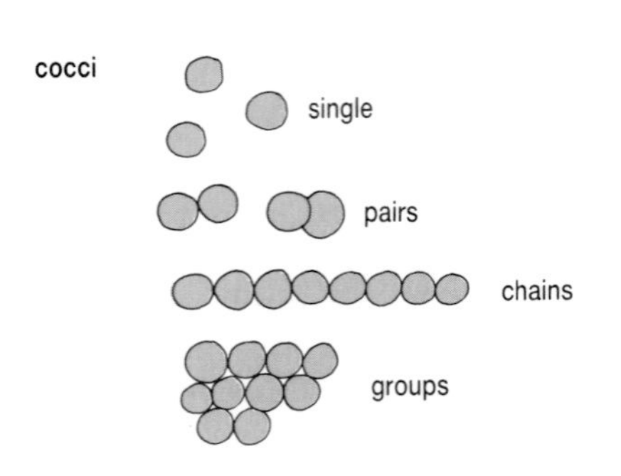

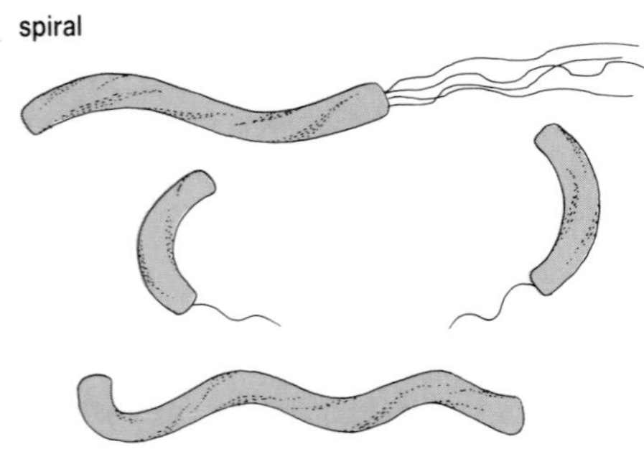

Fig 1.7 The shapes of bacteria.

Structure

Most species of bacteria have a typical shape and size, but some, for example *Rhizobium*, are **pleomorphic** which means they can take a variety of shapes. With the exception of the mycoplasmas, all possess a **cell wall**, which is very different to a typical plant cell wall. The wall gives the cell its shape and maintains the cell's structure. The outside of the cell may be coated with a **capsule**. Structures such as **flagella** project through the outer layers. The interior is membrane bounded cytoplasm with various inclusions and nucleic acid. Fig 1.8 illustrates the major features.

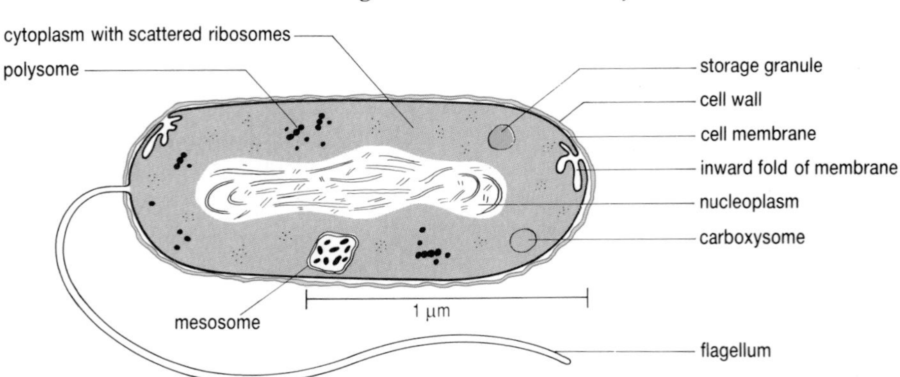

Fig 1.8 Bacterial structure.

The outer layers

The capsule is a barrier between the bacterium and its environment. It is often made of large 'gummy' polysaccharide molecules, which are secreted through the cell wall by the cell membrane. It is densest by the wall but thins out further away. Colonies of bacteria which make capsules appear glistening. The capsule can be washed off without affecting the survival of the bacteria and some strains do not make capsules at all. Capsules are difficult to stain so **negative staining** is used: the main part of the cell and the background is stained so the unstained capsule stands out against the background.

The capsule is very important to disease-causing bacteria as it increases their **virulence**, that is the ability to infect. This is because the capsule makes it difficult for white blood cells and antibodies to bind to the bacterium. In other bacteria the capsule may help to buffer the effects of changing water content in their environment. Some bacteria are now grown for the gummy material which has a range of uses.

The cell's shape and its resistance to osmotic changes depends on the cell wall. The cell wall is a tough, coarse meshwork of **murein** made of mucopeptides called peptidoglycans, synthesised by the cell membrane. The cell wall is not a barrier to the passage of molecules, even large enzymes and fragments of DNA can pass through.

Gram's Stain divides bacteria into two groups reflecting their mucopeptide content. Penicillin can only affect Gram positive bacteria.

Christian Gram developed a staining technique for bacteria in 1884 which is described in Fig 1.9. Bacteria are described as **Gram positive** if they take up the stain and **Gram negative** if they do not. The response to the stain is not caused by the different chemical composition of the cell walls, but it does divide bacteria into two groups whose cell walls are different. Gram positive bacteria cell walls have over 40 per cent peptidoglycans. Their thick wall contains other molecules too, many of which are antigenic, but there is little lipid or protein. The antibiotic **penicillin** affects Gram positive bacteria because it interferes with the synthesis of peptidoglycan by new bacterial cells. The molecules in such cells are not cross-linked properly so the wall is weaker than usual and when the cell gains water then osmosic stresses burst, or **lyse** it.

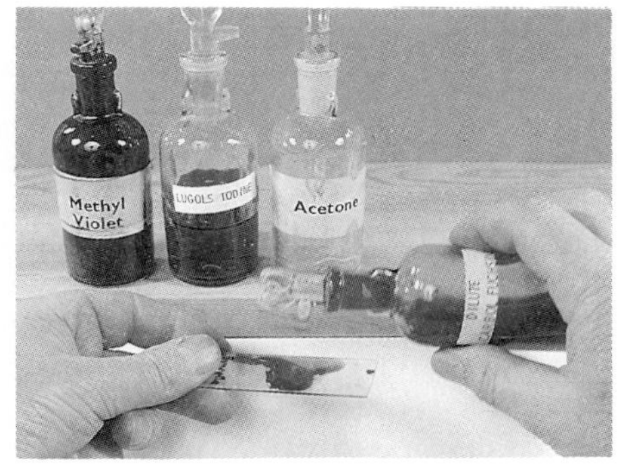

Fig 1.9 The procedure for Gram's staining technique. A smear is made of actively growing bacterial cells and heat fixed. The cells are stained with crystal violet and then dilute iodine. The slide is washed off with ethanol or propanone. Gram positive bacteria will retain the stain, but Gram negative bacteria will not. The gram negatives can be counterstained.

Gram negative bacteria have a two-layered cell wall which is much thinner than a Gram positive cell wall but more complex. The inner layer is rigid with little peptidoglycan: the outer layer is a membrane with proteins, lipids and lipopolysaccharides. Some Gram negative bacteria have projections on their surfaces called **pili** or fimbriae. These are cylindrical rods of a protein, pilin, that allow bacteria to link to each other, to other cells, and to the substrate. A sex pilus is a longer projection which allows a cell to donate nucleic acid to another cell.

The membrane inside the cell wall is similar to that in animal and plant cells. In some places it is extended, for example into thylakoids in photosynthetic bacteria, and into folds called **mesosomes** near the site of cell division. Folding gives the membrane a larger surface area, which is important since it is the site of most metabolic activity in the cell. It houses activities such as ATP-generating systems and enzymes for synthesising cell materials, as well as controlling the import and export of materials. Photosynthetic pigments and associated enzymes are also found on the surface or in the folds of the membrane.

The interior

The **cytoplasm** within a bacterium varies: some regions are denser than others but none is segregated from the rest. Ribosomes are scattered through the cytoplasm but some may be grouped as polysomes near the membrane. Cytoplasm holds materials for growth and maintenance and will also contain inclusions, particularly storage chemicals, for example glycogen, polyhydroxybutyric acid and lipids. The amounts of these depend on the environment and the cell's age. Some bacteria have inclusions concerned with CO_2 fixation or containing magnetite, a substance affected by the magnetic field, but little is known about these.

DNA, reproduction and genetic exchange

Bacterial cells have no separate nucleus although there is a region containing nucleic acids. Bacterial genes are located on a circular molecule, or **chromosome**, potentially able to encode 3000-5000 genes. About 1000 genes have been identified so far in *E. coli*. DNA duplication is not linked directly to cell division, so some bacteria may have more than one

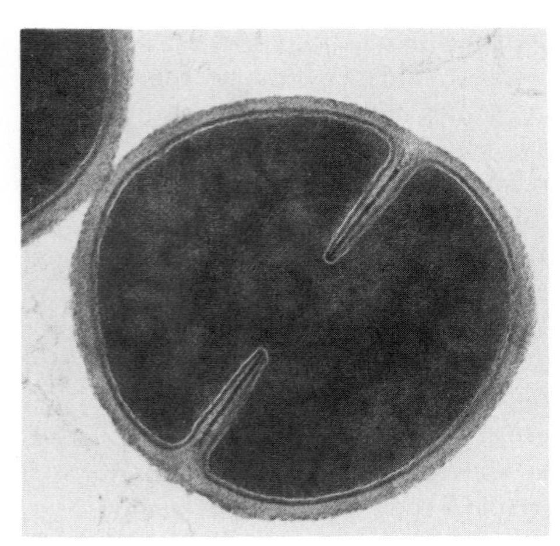

Fig 1.10 *Staphylococcus pyogenes aureus* undergoing binary fission. One cell has almost completed fission; when the developing cell walls meet in the centre the two halves will separate.

complement of DNA. When the DNA is duplicated the molecule is first attached to the cell membrane. Duplication starts there and continues by semi-conservative replication until it has got round to the beginning again. Bacteria reproduce asexually, by **binary fission**. When the cell divides, duplicated DNA is separated and the cell membrane folds inwards to form a double layer across the long axis of the cell. New cell wall layers are secreted within the membrane layers. This divides the cell into two smaller identical cells that may remain together or may separate. Dividing cells are shown in Fig 1.10. Binary fission is extremely rapid and will continue as long as the environment is favourable.

Bacteria do not reproduce sexually for there is no exchange of genetic material, but they can acquire new or different genes in other ways. Bacteria which have gained new genes are described as transformed bacteria. Extra DNA which has been duplicated can accumulate outside a bacterium in the capsule where it quickly breaks up into smaller fragments. These fragments may be absorbed by other bacteria and incorporated into their genome. Bacteria can also gain new DNA when certain viruses infect them. These viruses have reproduced inside other bacteria and carry fragments of bacterial DNA around with them.

Plasmids are a very important source of genetic material and about 70 per cent of bacteria carry plasmids. Plasmids are small self-replicating loops of DNA not linked to the main chromosome. The genes on the plasmid are not essential for growth but are often beneficial. Plasmids can be transferred between bacteria of the same species, but some can be transferred between different species. The transferred plasmid may be duplicated or it may be incorporated into the bacterium's chromosome. Some bacteria carry an F plasmid which allows direct genetic exchange between two cells through sex pili. Fig 1.11 shows bacteria linked by sex pili. The F plasmid passes through the pilus and takes some copies of donor genes with it.

Plasmids are extra loops of DNA carrying genes that may be advantageous but are not essential, and which can be passed from one bacterium to another.

Plasmids are useful to genetic engineers because they can be used to take genes into chosen cells such as bacteria, even genes not normally found in bacteria. The engineered genes will be duplicated and used by the new host bacterium. These transformed bacteria are used to make substances that are otherwise difficult to obtain, such as human growth hormone.

Nutrition

Bacteria have more ways of gaining energy and materials for growth than any other group of organisms. Their needs are similar to most animal and plant cells, but many synthesise compounds unique to bacteria and may use materials that no other organisms can cope with. Like all living things they need an energy source and carbon containing compounds. Energy may come from sunlight, from oxidation of organic or inorganic compounds, or very rarely from geothermal energy. It is said that all naturally occurring carbon compounds can be metabolised by bacteria. They may need other substances too, known as growth factors, to grow satisfactorily. Bacteria can be grouped according to how they gain energy.

Photoautotrophs

The green bacteria and the purple bacteria use light energy in photosynthesis to fix carbon dioxide and are described as photoautotrophs. They do not generate oxygen in photosynthesis because they use compounds such as hydrogen sulphide instead of water. Such bacteria release sulphur which may form deposits or it may be oxidised to sulphate ions. Not all the bacteria in this group are strict autotrophs; some purple bacteria can use light energy but obtain carbon from organic compounds. For example, *Rhodospirillum* can grow anaerobically using light to generate the ATP it needs by cyclic phosphorylation. This is then used to drive the breakdown of compounds such as acetate.

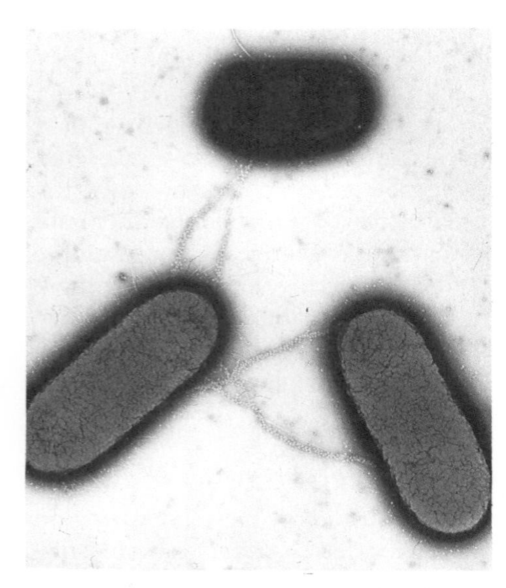

Fig 1.11 Three *E. coli* bacteria linked by F pili. One bacterium is donating DNA through the pili to the other two, transforming them.

Green bacteria are confined to oxygen-free environments, so they are restricted in where they can grow. They are found in shallow ponds with other bacteria which also thrive in the low oxygen conditions. Oxygen-releasing photosynthesisers, such as algae, live in the surface layers and harvest much of the light. Deeper down the purple and green bacteria harvest the longer wavelengths which have not been absorbed. The green and purple bacteria contain bacteriochlorophylls not chlorophyll *a*, and carotenoids for photosynthesis. The bacterial pigments enable them to absorb the longer wavelengths at the red and far red end of the spectrum. Fig 1.12 compares light absorption by pigments in bacteria and plants.

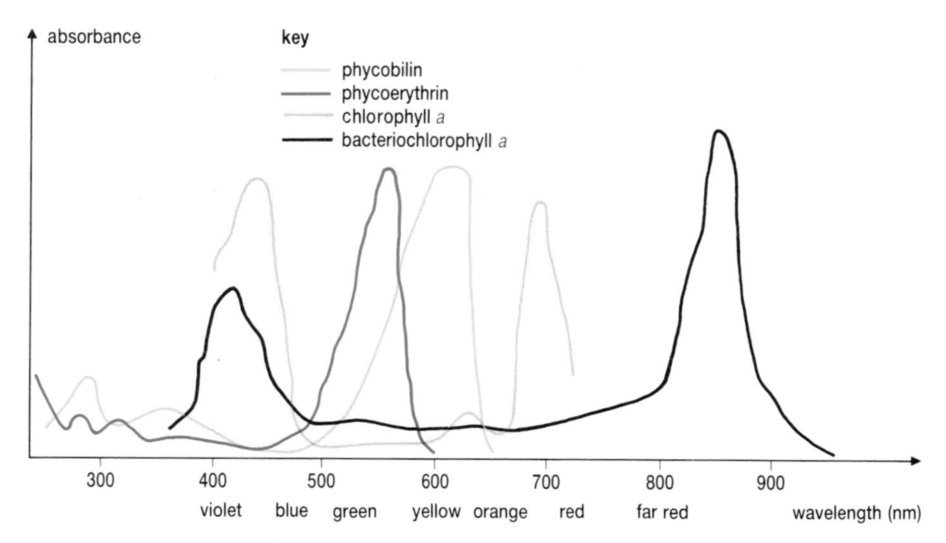

Fig 1.12 Absorption spectra of microbial pigments.

Chemoautotrophs

Photoautotrophs use inorganic light energy to fix CO_2. Chemoautotrophs use chemical oxidations instead.

Chemoautotrophs use energy from inorganic oxidations to drive ATP synthesis. Strict chemoautotrophs can be grown in the dark in a solution of mineral salts using carbon dioxide from the air. However some do need growth factors such as vitamins to grow. These bacteria are wide-spread in soils and water; the nitrifying bacteria in soil belong in this group. A variety of substances including hydrogen, nitrite ions, iron(II) compounds, hydrogen sulphide and others can be used as energy sources. These compounds are often produced by the activities of other bacteria and from straightforward geochemical reactions. For example, when iron pyrites is exposed to air in damp conditions, a chemical oxidation occurs which generates Fe^{2+} ions and acidic conditions. *Thiobacillus ferrooxidans* oxidises the ions to generate energy, and the end product reacts in acid conditions in a straightforward chemical reaction with more pyrites to generate more Fe^{2+} ions, and so the cycle goes on. It is thought that many of the naturally occurring pure iron ore and sulphur deposits are due to microbial activity.

Heterotrophs

Most bacteria are heterotrophs, metabolising organic carbon compounds to obtain energy and carbon for their own uses. They can use a wide range of compounds. Table 1.3 lists some of them. The bacteria metabolise these compounds to pyruvate and then carbon dioxide in ways similar to animals and plants. No matter how complex the nutrient molecule, it is always metabolised into a form that can enter the main pathways of respiration and be used to make ATP. Some bacteria are **aerobic,** using oxygen in respiration and releasing carbon dioxide. **Anaerobic** respirers use other molecules instead of oxygen, such as nitrate and sulphate ions. **Fermenters** fall within this category. They use the same pathway to break

Table 1.3
Some carbon-containing compounds metabolised by micro-organisms.

alcohols	cellulose
chitin	hydrocarbons
methane	methanol
monocarboxylic acids	polyethylene
polyurethanes	purines
pyramidines	steroids

down glucose and other compounds to pyruvate, but then pyruvate is metabolised in different ways, without using oxygen. The pyruvate may be converted into lactic acid, ethanol or a variety of other compounds. Fermenters can grow easily in a wide variety of habitats low in oxygen such as mud, animal gut and dead tissues.

Some bacteria can ferment if there is too little oxygen available for aerobic respiration. These are known as **facultative anaerobes**. The best known facultative anaerobe is in fact not a bacterium but a fungus *Saccharomyces cerevisae* (brewer's yeast). If an organism can use only one type of pathway then it is described as **obligate**. *Clostridium acetobutylicum*, for example, can only use anaerobic respiration. Often the products made by the respiration of one organism are used by others in the same ecosystem to form a mutually interdependent community.

Facultative anaerobes can switch to fermentation if there is insufficient oxygen for aerobic respiration.

Nitrogen Fixation

Prokaryotes are uniquely able to fix nitrogen from the atmosphere when there is no suitable nitrogen source such as ammonium ions available, though it results in slower growth. Most phototrophic bacteria and some other free-living Gram negative bacteria, such as *Azotobacter* and *Clostridia,* can fix up to 3 kg of available nitrogen per hectare per year, but there are endosymbiotic species which can fix 100 times more. Nitrogen, and other molecules such as hydrogen cyanide and ethyne, are reduced by nitrogenase enzyme to ammonia. The ammonia can then be used to make amino acids, bases and other nitrogenous compounds. The enzyme is very similar in all bacteria, being inactivated by oxygen and needing molybdenum for its binding site.

The main group of symbiotic nitrogen fixers are the *Rhizobia,* each species having a specific host in the Leguminosae. These aerobic heterotrophs can live in the soil, but if a suitable host grows nearby, the bacteria multiply and form a cellulose-coated infection peg which enters through young root hairs. They enter cortex cells and stimulate growth hormones which cause cell proliferation, resulting in the formation of nodules on the roots. The cells in the nodules house bacteria and a pigment similar to myoglobin, called **leghaemoglobin.** This is synthesised from components made by the cell and the bacterium and has a high affinity for oxygen, ensuring a supply for the bacteria in a low-oxygen environment. When the growth period finishes the bacteria disappear from cells and their materials are absorbed by the host plant.

There are other nitrogen fixing endosymbionts with other sorts of plants, and a nitrogen fixing gene has been transferred from one species to another raising the possibility of engineering nitrogen fixation into crop plants.

Mobility

Flexibility is restricted by the cell wall so many bacteria cannot move themselves but are instead moved passively in air currents and moisture films. Some are able to glide but the mechanisms are not properly understood. At least one group secretes surfactant chemicals affecting surface tension into the moisture film and this process propels them along. The most motile bacteria have **flagella,** shown in Fig 1.13, to swim through fluids. Bacteria may have a single flagellum, a few flagella in a tuft or have them all over their surface. The flagella may be much longer than the cells they propel. A bacterial flagellum comprises a bundle of entwined protein fibres erupting from the cell membrane, and has no definite membrane round it. It acts as a helical rotor, spinning round on its axis and propelling the bacterium along. Flagella do not work randomly, but all rotate in one direction at a time.

Fig 1.13 *Vibrio cholerae,* a water-borne bacterium, has a single flagellum located at the end of the cell.

Spores

The bacilli and clostridia all produce **endospores** as survival structures in adverse conditions. The spore develops inside a cell which is then broken open to release it. Spores are extremely resistant structures that can remain viable for many years and are able to withstand the effects of drought, ultra violet radiation, heat and many other stresses. Bacteria which produce spores can be a nuisance in many processes such as food preservation because although cells are killed, spores may not be unless extra care is taken. Many things can bring about spore germination, for example, heat shock, but often it isn't known what exactly triggers germination. Fig 1.14 shows a mature endospore.

Bacteria do not reproduce by spore production, a spore is a survival structure.

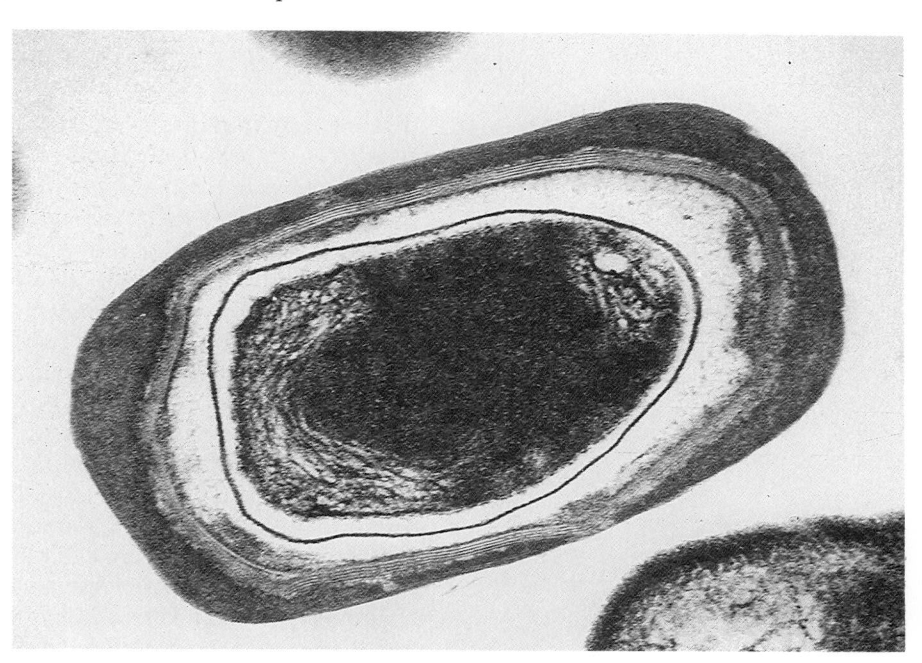

Fig 1.14 A spore of *Bacillus subtilis.* The thick outer coat surrounds an inner coat and membrane-bounded cortex.

QUESTIONS

1.4 What use is Gram's stain in the study of bacteria?

1.5 Why are some bacteria affected by penicillin but not others?

1.6 How do bacterial cell walls differ from the cell walls of flowering plants?

1.7 **(a)** What is meant by the terms 'heterotrophic nutrition' and 'chemoautotrophic nutrition' in bacteria?
(b) Are chemoautotrophs found in other groups of living organisms?

1.8 What do the terms facultative and obligate mean in bacterial nutrition?

1.9 Where and why do bacteria make endospores?

1.10 Sexual reproduction involving two individuals does not occur in bacteria. How do they carry out genetic exchange?

1.11 Examine Fig 1.12. Which wavelengths of light are absorbed more by green plants than bacteria?

1.12 For further reading investigate the role played by autotrophic organisms in the cycling of nitrogen and sulphur.

BRITAIN'S FIRST INDUSTRIAL DISEASE

Last century Bradford, in Yorkshire, had a prosperous textile industry. It was a major centre for spinning and weaving local and imported yarns. In the 1870s, Bradfordians were worried as more and more wool sorters were dying of blood poisoning. No-one knew what caused it, but when anthrax was identified by Robert Koch it was thought that wool sorter's disease could be anthrax. The disease was investigated, and it was shown that animals inoculated with blood from someone who had died of wool sorter's disease died of anthrax. It was discovered that the disease came with the wool, and that wools from some countries were more likely to bring about the illness than others.

A second form of the disease was discovered: a skin infection found in wool workers and workers using skins and hides in other industries. Other sufferers included wool workers' wives, railway porters and other people who only had fleeting contact with wool. Both forms of the disease were found to be caused by endospores in blood clots on the wool. The anthrax bacterium, which was common in many parts of the world, grows in the blood of an infected animal and the wool and hides had become infected during shearing or skinning. The spores were loosened during transport and processing and were inhaled or got into people's clothing causing infections.

In 1895 anthrax was declared Britain's first notifiable disease and the death of factory workers had to be reported. By 1905 local people had organised and financed a proper study. Methods to get rid of the endospores were investigated but the usual heat treatments damaged wool. By 1913 the government was worried enough to set up an inquiry to find a method of decontamination. There was an increase in cases during the next few years but by 1918 a treatment with formalin was found to be successful. The Government set up the Liverpool Disinfecting Station where all imported wool was treated. The effect on the disease was dramatic: there was a decline in cases, helped by improved general health and hygiene and the control of anthrax has greatly reduced the number of infected animals. Disinfection is now done in Bradford. Anthrax endospores are still present in some wools from some parts of the world, though, so workers in vulnerable trades still have to be particularly careful.

1.8 CYANOBACTERIA

Cyanobacteria, or blue-green bacteria, are similar to photosynthetic Gram negative bacteria though for years they were classified as algae. The name, blue-green bacteria comes from the colour of the cells which contain a range of photosynthetic pigments. The range of colours includes shades of red and yellow too, depending on the exact proportions of pigments.

Ecology and importance

Cyanobacteria are important photosynthesisers in aquatic ecosystems and are responsible for much of the productivity. Very few cyanobacteria need growth factors so they can grow in nutrient-poor waters and soils, as long as the temperature and light are adequate. Their photosynthesis fixes large amounts of CO_2 and they can also fix nitrogen when there is not enough suitable combined nitrogen. The fixed nitrogen can be used by other plants when the cyanobacteria die.

Some species are unicellular and float freely or are buoyed up by gas vacuoles in surface waters, others are filamentous and may be attached to plants or solid surfaces in water or moist places. They are found in seas, in freshwater, in warm alkaline springs, in soil moisture, as black streaks on the rocks on mountains, in microfissures in rocks in deserts and even in polar ice. When conditions are favourable they thrive in large numbers colouring the water. One type of blue-green bacterium, *Spirulina,* seen in Fig 1.15(b) has been eaten by people for many centuries. It grows well in salty alkaline lakes in Mexico and Chad where people harvest the long filaments and dry them in the sun to make a protein-rich food.

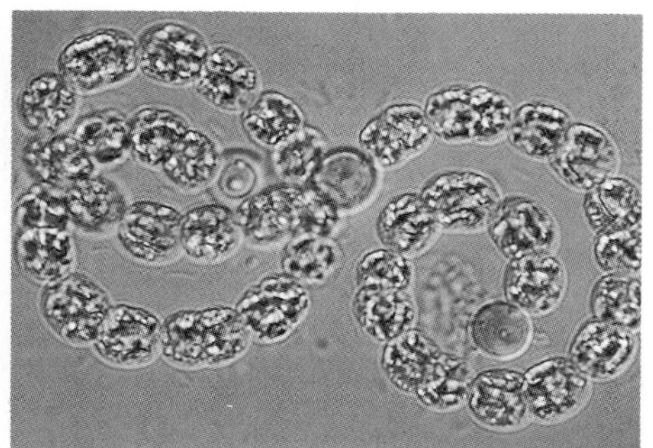
(a)

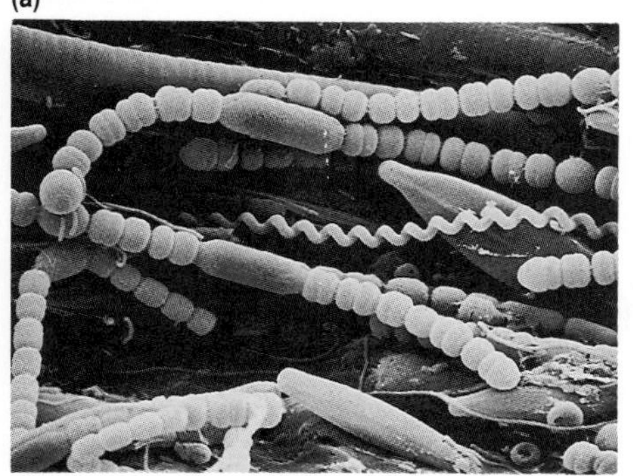
(b)

Fig 1.15 **(a)** A filament of *Anabaena circularis.* A small number of cells in the filament have become heterocysts.
(b) A scanning electron micrograph of cyanobacteria. *Anabaena* can be seen together with the corkscrew shape of *Spirulina* and the tubular *Microcoleus.*

Structure and Reproduction

The cells are rounded or elongated, and are often found in small clusters or long chains, as they are in the *Anabaena circularis* shown in Fig 1.15(a). A layer of mucilage may coat the outside of the cells making up a filament. Cyanobacteria reproduce asexually; some bud off new cells from older cells, others reproduce by binary fission. Filamentous forms may fragment into shorter lengths, followed by binary fission within the smaller chains. Internally their structure is similar to Gram negative bacteria, but they are exceptional among prokaryotes because they have photosynthetic pigments located in a series of coiled membranes, or **lamellae,** close to the outside edges of the cells. These membranes are not as complex as chloroplasts however.

Metabolism

All cyanobacteria contain chlorophyll *a* and other pigments such as **phycobilins,** which give a bluish colour, and a β-carotene. In addition some also have **phycoerythrins** which give a reddish tinge. These pigments allow the cells to harvest light in the range 550-700 nm, so they can utilise the greenish blue light in deeper water. Fig 1.12 illustrates the range of wavelengths these bacteria can harvest. The exact amount of each pigment synthesised is determined by the available wavelengths of light. This adaptation to the environment is called **chromatic adaptation** and allows the cells to exploit their environment to the full.

Chromatic adaptation allows blue-green bacteria to produce pigments to exploit the wavelengths of light available.

Photosynthesis in cyanobacteria is similar to that in land plants for the cells generate oxygen and fix CO_2 by the Calvin cycle and store glycogen. They do not have the same respiratory metabolism, though, and cannot use many organic compounds as energy sources, which means that they are obligate autotrophs.

If there are plenty of nitrate or ammonium ions available cyanobacteria do not fix nitrogen, but if supplies are short some cells become specialised for nitrogen fixation. These specialised cells are called **heterocysts.** They are thick-walled structures forming upto 10 per cent of cells, found singly or as a few clustered together in a chain of ordinary cells. Heterocysts can be seen interspersed with ordinary cells in Fig 1.15(a); an enlarged view is shown in Fig 1.16. The nitrogenase enzyme, which fixes nitrogen needs oxygen-free surroundings so ordinary cells with oxygen-generating photosynthetic processes going on are unsuitable for nitrogen fixation. A cell changes when it becomes a heterocyst, for example the lamellae with photosynthetic pigments are reorganised. The phycobilins are lost, and the cell can generate some ATP, but it cannot carry out the processes that produce oxygen and this factor makes the cell interior more suitable for nitrogenase activity. The heterocysts cannot fix CO_2 either, so they depend on neighbouring cells to provide them with metabolites through minute cytoplasmic bridges in the thick walls. Heterocysts cannot reproduce themselves, nor can they revert to become normal cells.

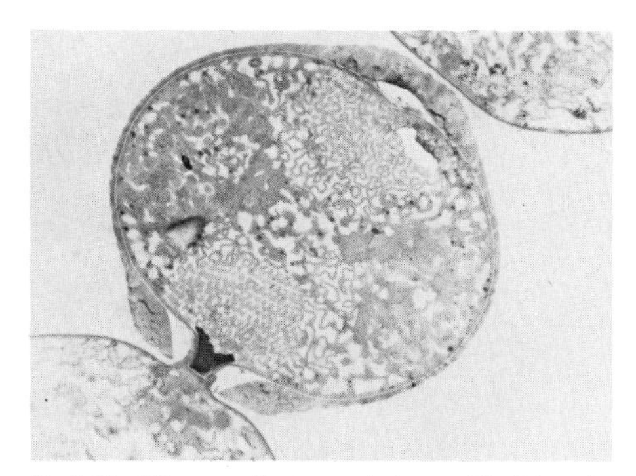
Fig 1.16 When a cell becomes heterocyst the internal structure alters; membranes with photosynthetic pigment are rearranged and capabilities are lost.

Nitrogen fixation can only occur in oxygen-free surroundings.

QUESTIONS

1.13 What is the function of a heterocyst?

1.14 Would a Gram stain help you investigate the cell wall of a blue-green bacterium? Explain your answer.

1.15 **(a)** What features do blue-green bacteria have in common with cells of flowering plants?
(b) Why are blue-green bacteria now grouped with bacteria?
(c) Give two reasons why blue-green algae are ecologically important.

1.9 PROTOZOA

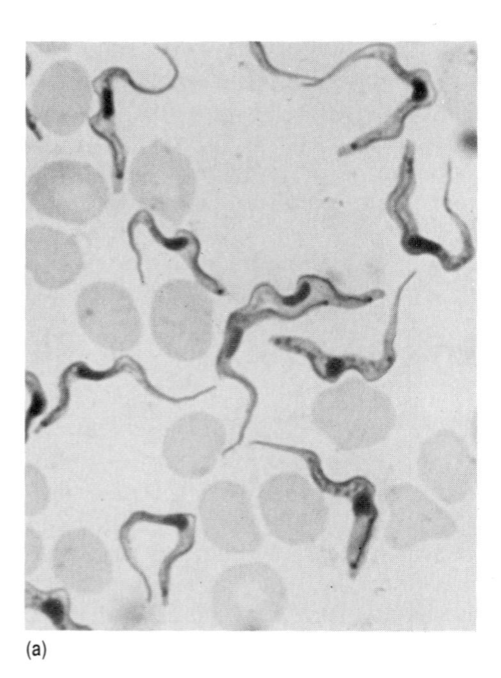

(a)

Protozoa are found in moist habitats, such as soil, freshwater ponds and rivers, the sea and animal body fluids. They tolerate a wide range of oxygen concentrations and pH. Many are herbivorous or feed on detritus, playing a large part in aquatic and soil food webs; others are predatory, feeding on bacteria and other protozoa. A few highly specialised protozoa are animal parasites, causing a number of important human and animal diseases including malaria, sleeping sickness and dysentery. This section can only give a brief survey of protozoal biology.

Movement

Protozoa are grouped by their means of movement; some exhibit a characteristic crawling movement, others move by using flagella or cilia to propel themselves through water or to create currents of water. Not all move about for some species are sedentary living attached to plants or detritus in water and these are classified by their similarities to other species. The **Rhizopoda** move in a way typified by, and named after, its most well known member, *Amoeba proteus*. Fig 1.17(b) shows an amoeba moving along a surface, which it does by extruding a portion of cell cytoplasm and membrane into an extension known as a **pseudopodium**. Changes in the state of the colloidal cytoplasm cause it to move gradually in the direction of the pseudopodium. The cell has no fixed shape, though some members of this group, for example the Radiolaria and some Foraminifera, form a rigid shell through which the pseudopodia extend.

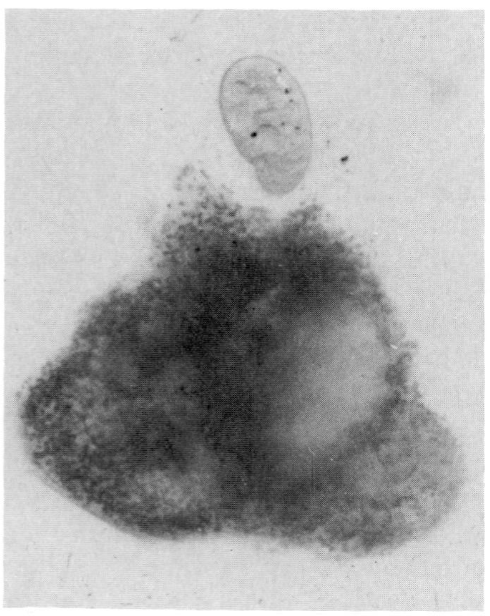

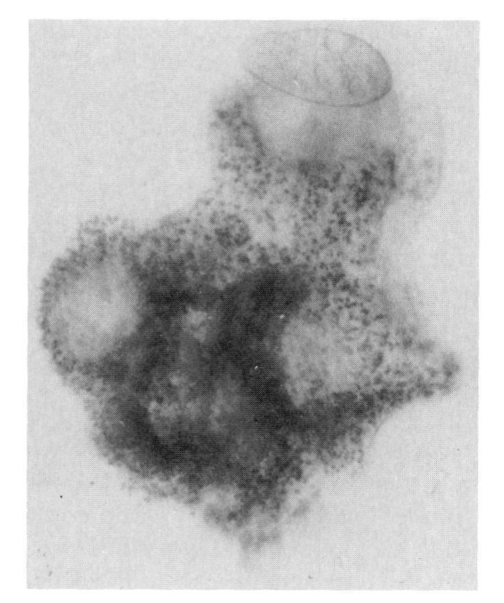

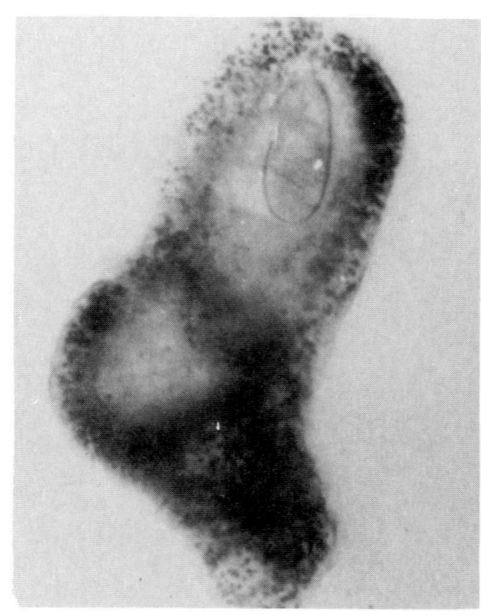

(b)

Fig 1.17 **(a)** Trypanosomes in a blood film; these move by using their flagella. The flagellum partially attached to the cell body by a membrane like a sail.
(b) Amoeba move by extending pseudopodia in the direction they wish to travel in and drawing the cell along. In this sequence an amoeba is using pseudopodia to engulf another protozoan (small and oval) as it moves along.

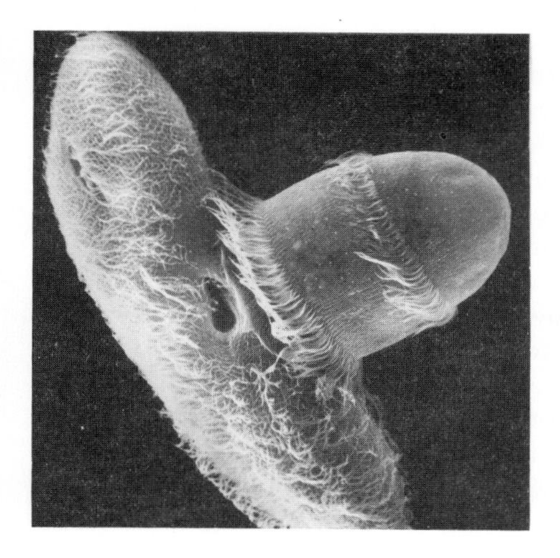

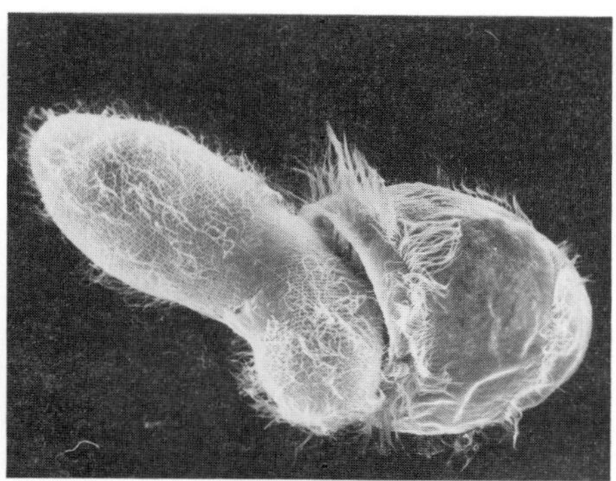

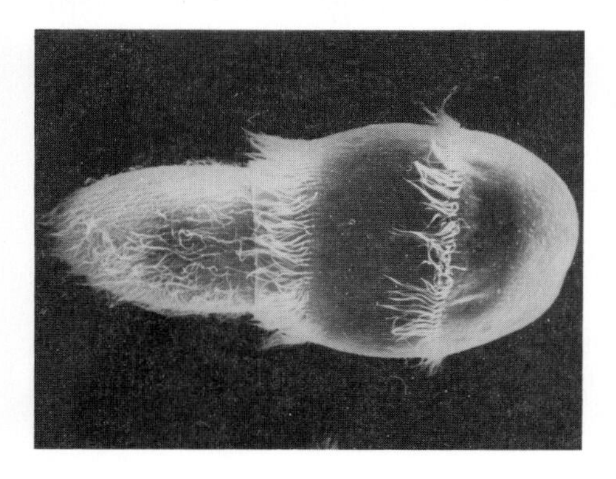

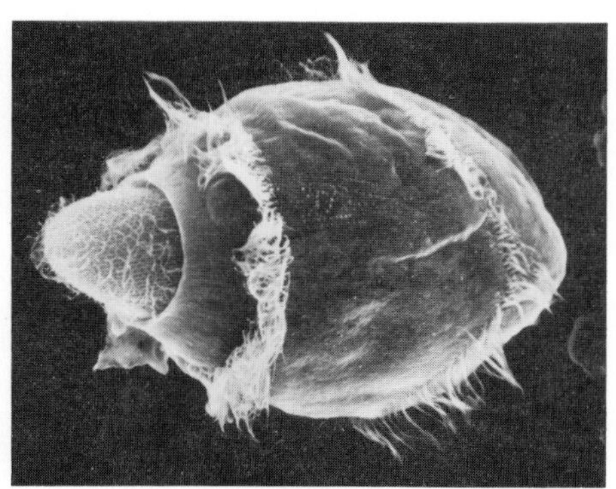

Fig 1.18 One predatory ciliate, *Didinium*, which is shaped like an egg cup, ingesting another ciliate, *Paramecium*.

The **Mastigophora** are protozoa that move using one or more flagella, seen in Fig 1.17(a). Many of them are parasitic, such as *Trypanosomas* which causes human sleeping sickness. The trypanosomes have a single flagellum that runs forwards along the length of the cell, attached by a fine membrane, and projects beyond the cell body. As the flagellum moves it moves the membrane too and propels the organism through the blood.

The **Ciliata** are a familiar group of protozoa and they can often be seen in hay infusions and pond water. In particular, *Paramecium caudatum* can frequently be seen pursuing its spiral path through water. Movement is by the beating of rows of cilia over the whole surface of the cell. The action is synchronised and the organisms can change direction quickly. Some ciliates are fixed to their substrate and use cilia to generate a current of water to bring food particles into their feeding structures. Fig 1.18 shows a predatory ciliate, *Didinium*, ingesting another, *Paramecium*.

Sporozoa are a mixed group. They are all parasitic, for example *Plasmodium falciparum* which causes malaria, and may move by gliding or flexing.

Physiology

Protozoa are eukaryotic and much of their activity is similar to that of animal and plant cells. The difference is that the protozoan cell has to carry out all functions of the organism. Protozoa are large, ranging from 5 μm to 2 μm bounded by a tough cell membrane, or **pellicle**. All protozoa are heterotrophic, but they feed in different ways. Parasitic protozoa absorb soluble nutrients from surrounding fluids, whereas most of the rest of the group ingest particulate material. In some protozoa movement and feeding are inextricably linked together. For example, *Paramecia* have a structure, a spiral groove that extends part way round the body to a **cytostome**, leading into the interior of the cell. As the protozoan swims through water, a current of water bearing detritus fragments flows into the groove to the tip where a vacuole encloses the particles. Similarly, as amoeboid protozoa move about, the pseudopodia encounter particles which are engulfed in a vacuole formed by extensions of the cytoplasm round the particle. Enzymes secreted into the vacuole break down materials and anything that cannot be absorbed is egested as the organism moves on.

Freshwater protozoa are subject to osmotic stresses as water enters the cell by osmosis. Osmoregulation is by **contractile vacuoles** using respiratory energy. Surplus water enters small vacuoles which feed into larger vacuoles at certain spots in the cell. When these are filled they discharge at the cell surface. The rate at which vacuoles form depends on the rate that water enters the cell.

Most protozoa avoid unfavourable environmental conditions simply by moving away from unfavourable stimuli. However, if conditions become too harsh some protozoa can form dormant cysts which may survive dessication or salinity. The cyst of *Acanthamoeba*, for example, has a very thick wall and inside it the organism becomes slightly dehydrated with a reduced metabolism. When conditions are more favourable the organism leaves its cyst.

Reproduction

Most protozoa reproduce by binary fission; the cell nuclei and organelles are duplicated, then they, and the cytoplasm, are divided between two daughter cells. The process is quick, generating large populations rapidly. Some parasitic species undergo multiple fission which generates very large populations. Some protozoa, such as *Amoeba proteus* and *Trypanosomas*, can only reproduce asexually, but many protozoa cannot continue reproducing this way indefinitely. In these species genetic exchange is needed to prolong the life of the cell line.

The process of sexual reproduction varies between species, and it is impossible to define a general pattern. Some species are haploid, others are diploid, and some have multiple nuclei. All sexually reproducing species do, at some point, form special nuclei, though not necessarily separate gametes. During sexual reproduction nuclei carrying genetic material are exchanged and fused. Usually the exchange is between two individuals, though sometimes only one individual is involved.

It is beyond the scope of this book to go into details of the many varieties of sexual reproduction, but the following brief resume of the sex life of the ciliate *Paramecium aurelia*, shows how complex the process of reproduction can be. This species has three nuclei. One nucleus, the macronucleus, is larger than the others, and its functions include controlling metabolism and cell division, which is by binary fission. The smaller micronuclei are involved in the exchange of genetic material. Sexual reproduction, called **conjugation**, usually involves a partial fusion of two cells of a particular mating type.

Step 1. Each micronucleus divides by meiosis to produce a total of eight haploid nuclei. Seven disintegrate and the eighth nucleus divides mitotically to make two daughters.

Step 2. One of these daughters is exchanged. The exchanged nucleus fuses with the host nucleus to form a new nucleus with genetic material from two individuals. The macronucleus disintegrates.

Step 3. The cells separate.

Step 4. The fused nucleus multiplies mitotically and generates new micro- and macronuclei which are allocated in the next binary fission between the daughter cells.

Notice that there is no increase in cell numbers until the final binary fission, but the individuals are genetically different.

QUESTIONS

1.16 What uses do ciliates put their cilia to?

1.17 How does the structure of a protozoan such as *Amoeba proteus* differ from that of a bacterium?

1.10 EUGLENOIDS

Euglenoids are found in a variety of moist, light habitats. In sunlight they are photosynthetic and almost autotrophic but use vitamins B_1 and B_{12} as growth factors. In the dark they become heterotrophic, absorbing soluble nutrients from their surroundings. There are some entirely heterotrophic euglenoids which lack photosynthetic pigments. Some of these species absorb soluble molecules but others have specific structures for ingesting food.

Structure

Euglenoids have some algal features and some protozoal features. All Euglenoids, typified by *Euglena gracilis*, have a long narrow cell shape. There is no cell wall, unlike plant cells, instead the outer membrane is a thickened pellicle like that of protozoa and this gives a consistent shape to the cell. There are two flagella seated in a cavity at one end of the cell (see Fig 1.19(a)). One is long and used to propel the cell; the other doesn't extend out of the cavity but seems to be used to help the cell orientate itself to obtain maximum sunlight. Near the flagella base is a specialised structure, the red **eyespot**, which contains light-sensitive carotenoids. This is used, along with the short flagellum, in orientation. Photosynthesis is carried out in self-replicating chloroplasts arranged round the central region of the cell. Though they are large they are simpler in structure than those in plant cells. Chlorophyll *a* and *b* are used in photosynthesis but, instead of starch, β 1,3 glucan is stored. Some species have a red pigment, that migrates to the surface of the cell in very strong light, which is thought to protect against high light intensity.

flagellum
eyespot
second flagellum
contractile vacuole
chloroplast
nucleus
granule of β1,3 glucan
mitochondrion
10 μm

(a)

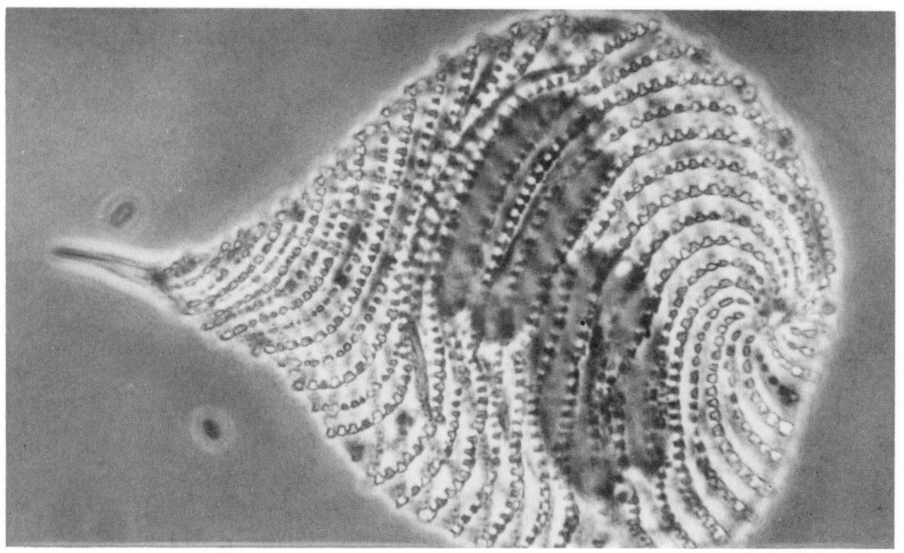

(b)

Fig 1.19 (a) Generalised euglenoid structure.
(b) Photomicrograph of *Euglena spirogyra*.

Reproduction

Sexual reproduction does not seem to occur in the Euglenoids. They reproduce by binary fission with division taking place down the length of the cell after the organelles have duplicated.

1.11 UNICELLULAR GREEN ALGAE

Unicellular green algae, or Chlorophyta, are found in fresh- and seawater as plankton. They are the commonest photosynthesisers in aquatic ecosystems and contribute greatly to productivity. They are also found in soils and other moist environments; the green encrustation of *Pleurococcus* on the barks of trees and damp walls is a familiar sight. Many tolerate extreme habitats; species can even be found living in melted snow. Some species form symbiotic relationships with other organisms such as sponges, protozoa and fungi. Some **lichens** are mutualistic relationships between fungi and algae.

Calvin used a common unicellular green alga, *Chlorella*, in the investigations which unravelled the processes of carbon fixation in photosynthesis. Potentially unicellular green algae could be grown as sources of single-cell protein and other useful substances, or to extract metal ions from water. Green algae can take up metal ions from low concentrations in surrounding water and this could be exploited to purify contaminated water or to extract precious metals from wastes.

Structure and physiology

The unicellular green algae have cellulose cell walls, pectin and other materials though desmids and diatoms, for example, have silicate shells not cellulose walls. The internal structures vary from species to species but *Chlamydomonas*, shown in Fig 1.20, is fairly typical. The single curved chloroplast, with chlorophyll *a* and *b*, and sometimes a protective red pigment, occupies most of the cell. The organism has a light-sensitive carotene area and orientates itself to receive maximum sunlight. Other organelles are found in the area enclosed by the chloroplast. Movement towards a light source is brought about by two flagella, projecting through the cell wall, which propel the organism through water. Unusually for a cell with a cell wall which supports it, it has contractile vacuoles to regulate the water content.

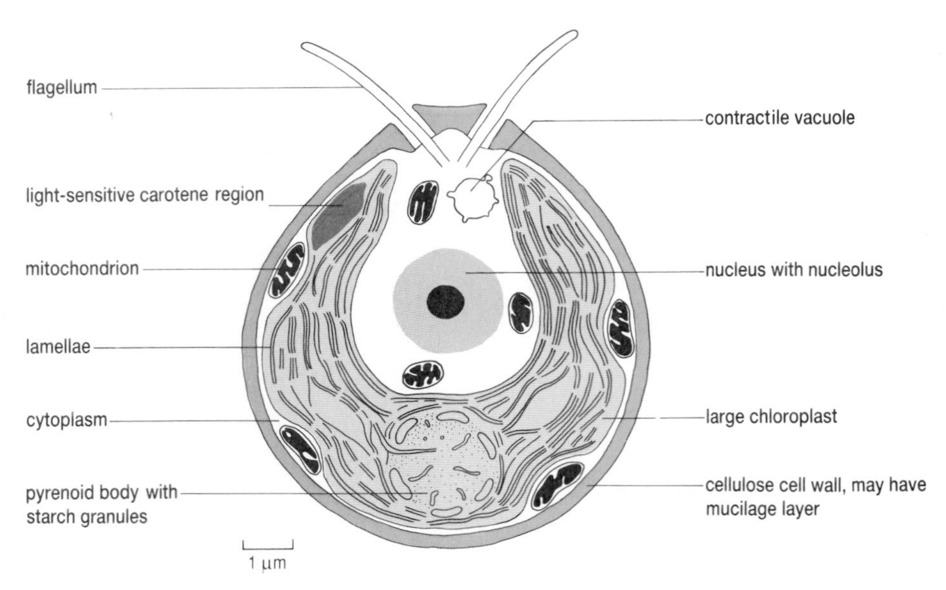

Fig 1.20 Diagrammatic structure of *Chlamydomonas*.

Reproduction

Chlamydomonas reproduces both asexually and sexually. Asexual reproduction is different to binary fission as it results in four or more daughter cells instead of two. The organism withdraws its flagella, organelles are duplicated then shared with cytoplasm to produce a number of cells within the original cell wall. The daughter cells secrete new cell walls and flagella and are then liberated from the parental cell wall. If the surroundings are particularly hostile the cells can remain within the parental wall until conditions improve.

Sexual reproduction occurs less often. The cells lose their external wall and become flagellate gametes that can fuse with other similar cells. The resulting zygote develops a thick wall and then divides meiotically to produce four daughter cells.

QUESTIONS

1.18 At various times Euglenoids have been grouped with protozoa and with algae. Give three features which link them to protozoa and three features which they have in common with algae.

1.19 What features do the unicellular green algae possess that Euglenoids do not?

1.20 *Pleurococcus*, a unicellular green alga confined to moist places, often grows more densely on the northern side of a tree trunk. Outline an investigation you could carry out to determine whether moisture content of the air around the bark is an important factor in the distribution of *Pleurococcus*.

EUTROPHICATION

Many rivers, lakes and seas receive far more nitrates and phosphates than they used to. These come from two main sources. One is agricultural fertiliser which is leached out of soil by rain and drains into watercourses. It is estimated that half of the average 150 kg of inorganic nitrate put on each hectare of land each year is not taken up by plants but is carried as dissolved nitrate to the water table. The other source is discharges from sewage works which carry large quantities of these ions.

Planktonic blue-green bacteria and unicellular algae thrive in the enriched waters, multiplying rapidly. The vast number of cells produced leads eventually to the death of many of the other organisms in the water. Oxygen generated in photosynthesis buoys up the cells in the surface waters forming a dense 'soup' that reduces the amount of light reaching plants in deeper water. When the cells die they sink into deeper, colder waters and provide a rich source of nutrients for heterotrophic bacteria and other decomposers. The activity of these organisms quickly depletes the oxygen available in the water and causes the death of many oxygen-dependent creatures such as fish, leaving only the anaerobic bacteria. Their anaerobic respiration produces hydrogen sulphide, acids and amines which make the waters even more unsuitable for most aquatic organisms, and also make the water and mud smell very unpleasant. This process is called **eutrophication**.

1.12 FUNGI

Fungi are common in soil and water. They are aerobic, usually confined to surfaces where there are high oxygen concentrations, though yeasts are more flexible in their oxygen requirement. Fungi can grow in substances with very low water potentials, such as wood, and are very tolerant of acidity. They are heterotrophs which obtain nutrients by secreting enzymes into the immediate environment and absorbing soluble products through the cell wall. Many live on dead organic material and are described as **necrotrophic**. They are among the first organisms to degrade dead material, releasing components that can be used by other organisms in an ecosystem. The fungi produce a very wide range of enzymes, including some found in few other organisms, for example cellulases and lignases. They therefore play a key role in the cycling of organic matter. **Biotrophs** are fungi which are adapted for a parasitic life in animals and plants. Fungi spread as spores which are produced in large numbers. Spores are released into air or water and disperse widely, some travelling hundreds of kilometres. Warm moist conditions are ideal for their germination and any organic material left in these conditions will quickly develop a fungal growth.

Necrotrophs degrade dead organic material, biotrophs obtain nutrients from living cells.

Yeasts have long been used commercially for brewing and baking, but other fungal species are now being exploited on a large scale because they can produce a range of enzymes and other useful substances. For example some fungi secrete antibiotics.

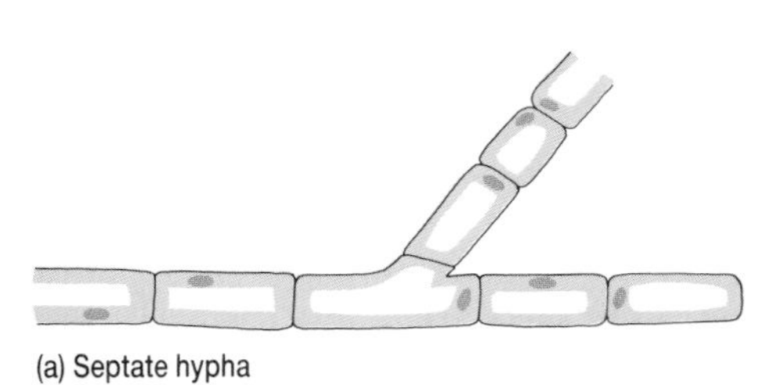

(a) Septate hypha

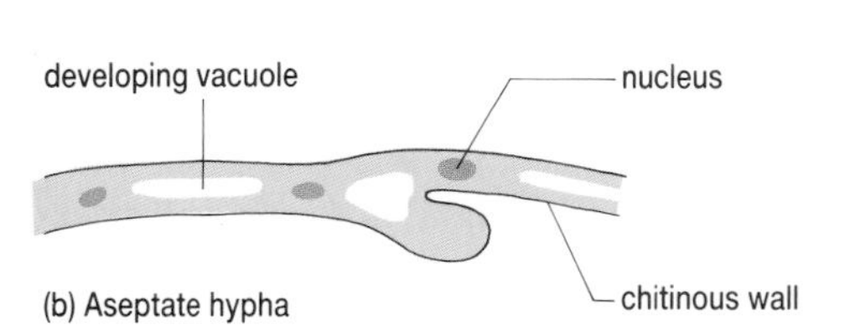

(b) Aseptate hypha

Fig 1.21 The structure of hyphae.

Structure

The least complex in the group are protozoa-like, but most are filamentous multicellular structures. The filaments are called **hyphae** which together make a mass called a **mycelium**. In some groups hyphae have cross walls, and are said to be **septate**, the septa dividing the hyphae into cells. In others the cytoplasm has many nuclei scattered along its length. This sort of multinucleate cell is called a **coenocyte**. The structure of hyphae can be seen in Fig 1.21.

Each mycelium starts with the germination of a single spore which grows a hypha. The hypha branches, eventually forming a circular mat as

Fig 1.22 Fungal hyphae grow in all directions, providing there are enough nutrients to support growth. This results in a circular mycelium.

in Fig 1.22, extending over the surface of the substrate. If oxygen concentrations are high enough hyphae will penetrate into the substrate. There is no limit to the size of the mycelium; it will grow wherever there are enough nutrients. In large mycelia the oldest parts, which are in the centre, may die.

Many fungi are able to translocate nutrients from one part of the mycelium to another, supporting hyphae growing over stone or unfavourable substrates. Most fungi survive difficult conditions as spores, but some species form survival structures. For example *Botrytis*, a fungus affecting crops, forms **sclerotia** of tightly packed hyphae, with thick outer walls. These can survive in the soil for years until a suitable plant grows there again.

Reproduction

All fungi can grow from fragments of the mycelium which will form a new mycelium. They can also produce large numbers of sexual and asexual spores from sporing structures. Fungi are grouped into three groups according to their method of sexual spore production. Unfortunately, every part of the reproductive process involves difficult names for quite simple processes – remembering the names can be far more difficult than understanding the processes!

Oomycetes and Zygomycetes

These fungi have the simplest structures. Oomycetes are often aquatic; zygomycetes are generally terrestrial but there are exceptions. *Saprolegnia*, the water mould, *Rhizopus*, or black bread mould and *Mucor*, or pin mould, are members of these groups which are familiar to many people. Both groups grow extensive mycelia which produce asexual spores within a few days. Spores are produced in chambers, **sporangia**, at the tips of hyphae extending beyond the mycelial mass. These are easy to see in terrestrial species as they project above the mycelium. Inside the sporangium a mass of sporangiospores are produced by mitosis which are dispersed into the air. Aquatic fungi such as *Saprolegnia* make flagellate spores, called **zoospores**, that swim away.

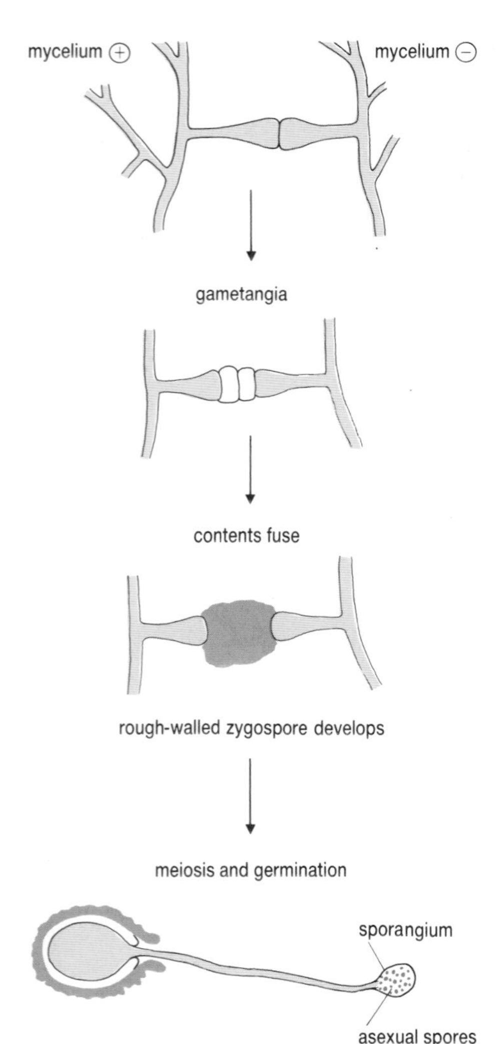

Fig 1.23 Sexual spore production in *Rhizopus*.

Sexual **zygospores** are produced when the environment becomes harsher. The aquatic forms produce male and female gametes which fuse; the process in a terrestrial form is slightly more complex. Zygospore production in *Rhizopus* is shown in Fig 1.23. In some species sexual reproduction can only take place if two different mating strains grow next to each other, but it is difficult for mycologists to distinguish any differences between them. These fungi are described as **heterothallic**.

Ascomycetes

These are common and can be seen during most of the year. The group includes yeasts, black and green moulds, powdery mildews, morels and the truffle. They cause many problems, from the rotting of cloth to apple scab, ringworm, wheat diseases and Dutch Elm Disease. Some species are grown commercially to make products such as penicillin, citric acid and other useful substances. During active growth the septate mycelium grows hyphae called **conidiophores** which produce asexual spores. The spores, called **conidiospores**, are budded off the end by mitosis; some can be seen in Fig 1.24. They form long often brightly coloured chains and are dispersed in air currents.

The sexual spores are **ascospores**, produced in an **ascus**, a sac at the end of a special hypha with binucleate cells. The processes that bring together nuclei are complicated and vary from species to species. Ascospore formation is shown in Fig 1.25. The spores are released when the ascus ruptures.

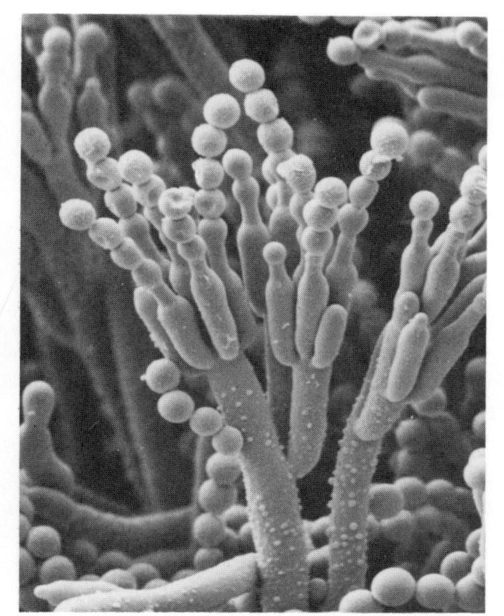

Fig 1.24 Chains of conidiospores carried on the end of a special hypha (*Penicillium camembert*).

One group of Ascomycetes is different, these are the yeasts. The many yeast species do not grow as mycelia but exist as single cells reproducing by budding. Yeasts are described in more detail at the end of this section.

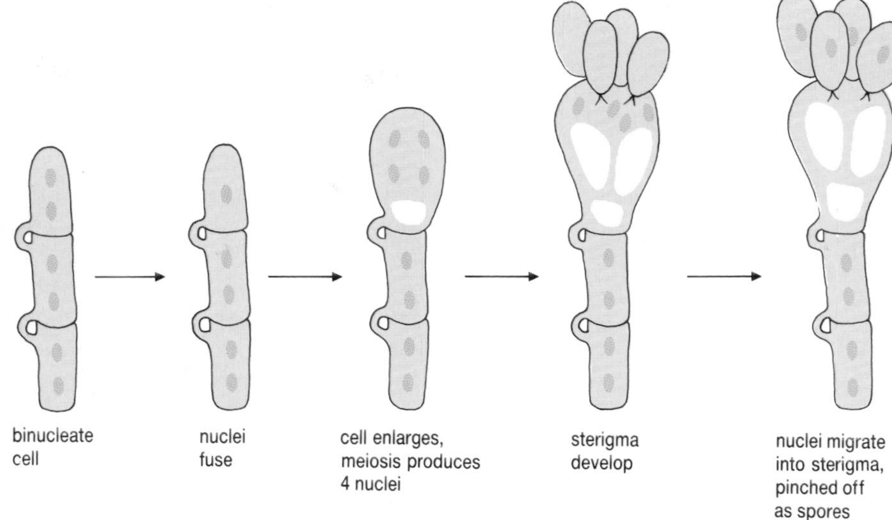

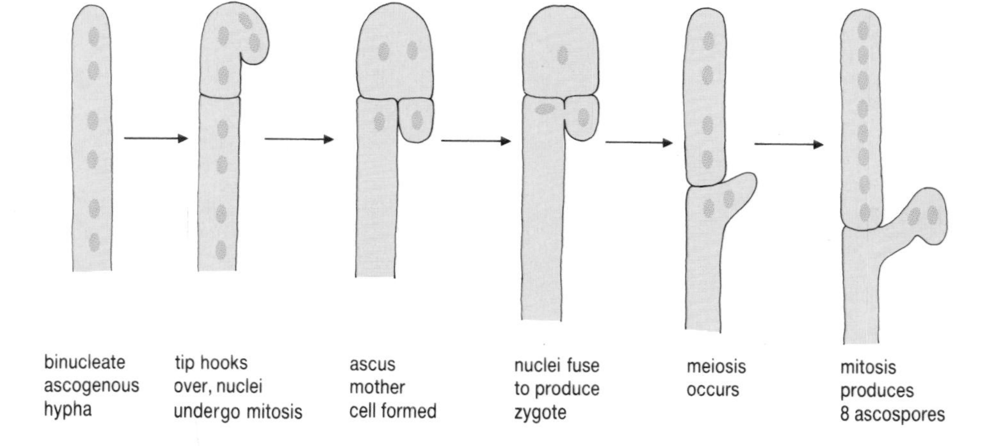

Fig 1.25 Formation of ascospores and basidiospores.

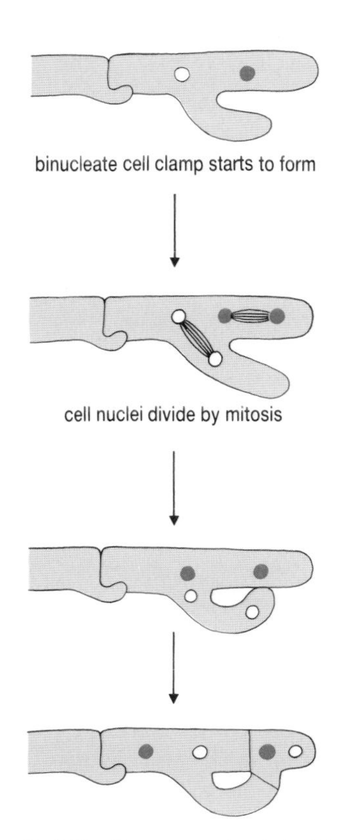

Fig 1.26 Clamp connections in basidiomycetes.

Basidiomycetes

The Basidiomycetes make the large fruiting bodies called mushrooms and toadstools though the fruiting body in many species may be microscopically small. The group includes bracket fungi, bird's nest fungi, smuts and rusts. They seldom make asexual spores, but produce sexual **basidiospores** in a fruiting body. Most of the year the fungus grows through its substrate – soil, wood or vegetation – as an extensive mycelium. The hyphae are separate, but binucleate cells are produced in some species when the fungus reproduces sexually to make the fruiting body. As cells divide, a structure called a **clamp connection**, shown in Fig 1.26, ensures that each daughter cell gets one of each of the daughter nuclei. Inside the fruiting structure hyphae grow into layers of club-shaped **basidia** each forming four haploid **basidiospores** by the process shown in Fig 1.25.

Fungi imperfecti

Fungi imperfecti is a 'catch-all' group of species whose method of sexual reproduction is unknown and which do not have enough other features to assign them satisfactorily to one of the groups.

YEASTS

Yeasts are found everywhere, on the skins of fruit and vegetables, in dust, dung, soil, water, milk and on animal mucous membranes. The earliest historical records mention the use of yeasts to ferment fruit and grain sugars to make alcoholic drinks and as leaveners in bread. The term 'yeast' is really a non-specific name for a whole group of organisms which use sugars and produce carbon dioxide and ethanol. The yeasts, in fact, fall into two groups. One group uses aerobic respiration and does not make ethanol. The presence of oxygen stimulates the breakdown of pyruvate and oxidising sugars completely to carbon dioxide. These yeasts grow very rapidly and some, for example *Hansenula*, are important spoilage organisms. The members of the other group, which includes *S. cerevisiae*, baker's yeast, and *S. carlsbergensis*, a brewer's yeast, can tolerate low oxygen environments. They break down glucose to pyruvate to obtain energy then convert it to ethanol as long as there is plenty of free glucose in their environment. If the glucose levels fall they switch to making ethanol into carbon dioxide. These products, carbon dioxide which causes bread to rise, and ethanol which gives alcoholic drinks their potency, are the bases of vast industries. Surplus yeasts from brewing are used as a source of vitamin B complex and as animal feed.

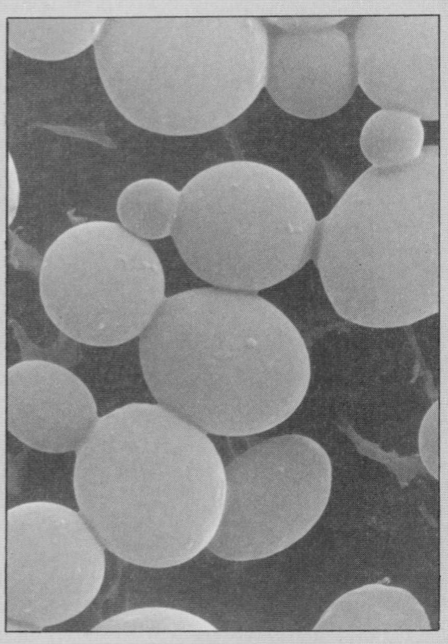

Fig 1.27 Yeast cells reproduce by budding smaller cells, these may remain attached together forming clumps.

When a sample of baker's yeast is examined it appears as ovoid cells surrounded by a thick cell wall. You can see a cluster of yeast cells in Fig 1.27. Some cells appear to have smaller cells attached to them. These are daughter cells, produced by budding, which have remained attached forming clumps. Yeasts survive unsuitable conditions as dormant ascospores, produced from the fusion of two yeast cells. The nuclei fuse then undergo meiosis to make haploid spores within the original cell wall. However, a nucleus may undergo meiosis followed by the fusion of pairs of nuclei to give diploid zygotes which bud off, so a wild population of yeasts may have both haploid and diploid cells.

There are hundreds of yeast strains each with its own characteristics and abilities. Different ones are used for different products, but manufacturers are trying to improve their strains. It is difficult to breed improved varieties as yeasts do not mate randomly; only with cells of the same strain making 'cross breeding' very difficult. Commercial strains are often genetically abnormal with multiple sets of chromosomes or with missing chromosomes, this too makes it difficult to do genetic crosses with them.

Yeast users would like yeasts which can tolerate high ethanol concentrations and strains which can make their own amylase. Yeasts cannot degrade starch into simpler sugars – brewers have to 'malt' grain starches into sugars before they can be used in brewing. This is done by either allowing the grain to sprout so it produces its own enzymes to degrade the starch, which takes days, or by adding commercial enzymes. Yeasts which could degrade starches would remove an expensive process. As people are becoming more aware of the health and social problems caused by alcohol, the market for low alcohol beers is expanding, so brewers are also looking for strains which do not make much ethanol but still make all the flavour chemicals responsible for the taste of beer. Some brewers would like a yeast strain which can degrade insoluble β-glucans, made from barley starch in the process of breaking down into sugars, because they tend to clog the brewery equipment. Efforts are being made to engineer desirable genes into yeast cells using plasmids but these are not always successful. As brewer's yeast is made into yeast extract spreads for people, any engineered strains must be absolutely safe for the consumer. A new line of research is to use yeasts as organisms which can be engineered to synthesise a variety of products in the same way as bacteria.

QUESTIONS

1.21 In industrial processes using *Aspergillus* the inoculating material put into the substrate is often described as a 'conidial suspension'. What do you think this means?

1.22 What is the importance of extracellular enzyme production in the life of a fungus?

1.23 Fungi readily grow on cotton fabric, which is over 90 per cent cellulose, in warm, moist climates. If the fabric is tested it is found to have lost much of its strength. Why do you think this loss of strength occurs?

1.24 What environmental factors limit the growth of a fungal mycelium?

1.25 If your house is found to suffer a dry rot fungus infection (*Serpula lachrymans*) you are recommended to remove infected wood and surrounding wood and to burn both. Why does the wood round the infection have to be burned as well?

1.26 When are the different types of fungal spores produced?

1.27 Briefly describe the difference between asexual spore production and sexual spore production in the fungus *Rhizopus*.

1.13 VIRUSES

Viruses are obligate intracellular parasites.

The first time a disease was identified as viral was in 1892, but the causative agents were not isolated for many years because of their small size. Viruses are obligate intracellular parasites which cannot multiply or be active outside their host cell. Outside cells, viruses are **virions** made of a protein coat wrapped round nucleic acid but have no other cell components. Virions can infect susceptible cells and divert normal cell activity into virus-directed activity. This disruption of cell activity almost always has harmful effects on the host. Most viruses are known to cause harmful effects on host cell metabolism, but some are temperate and do not cause damage immediately. Viruses are specialised for growth within a particular sort of host cell and there are virus parasites of almost all living things.

Structure

Viruses are very small, ranging from about 20 nm to 400 nm in size. Each virus has a simple structure, namely a molecule of nucleic acid as its core, arranged within a protein coat, or **capsid**. The aggregate of nucleic acid and protein coat is the **nucleocapsid**. Some types of viruses acquire a membrane layer around the outside of the capsid, called an **envelope**, as they emerge through the host cell membrane. Fig 1.28 shows enveloped viruses emerging from a cell. The nucleic acid is not arranged round proteins as in other organisms and may not even be DNA. Some viruses have double-stranded DNA as the genome, but others have single-stranded DNA, single-stranded RNA and even double-stranded RNA. The genome of the smallest virus, øX174, is about 10 genes but the nucleic acid in larger viruses is large enough to encode several hundred genes.

The capsid is made up of identical protein sub-units called **capsomeres** which link together spontaneously in regular arrays to form geometrical structures. Some viruses have twenty-sided (icosahedral) capsids, others may be helical. Fig 1.29 shows some of the commoner arrangements. The capsid may be made of more than one type of protein; the apices may have different proteins from the faces.

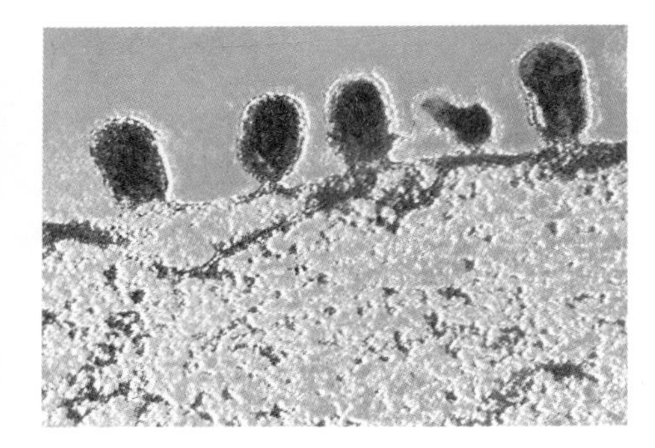

Fig 1.28 Influenza virus particles leaving the host cell. They acquire an outer coating of membrane as they pass through the host membrane.

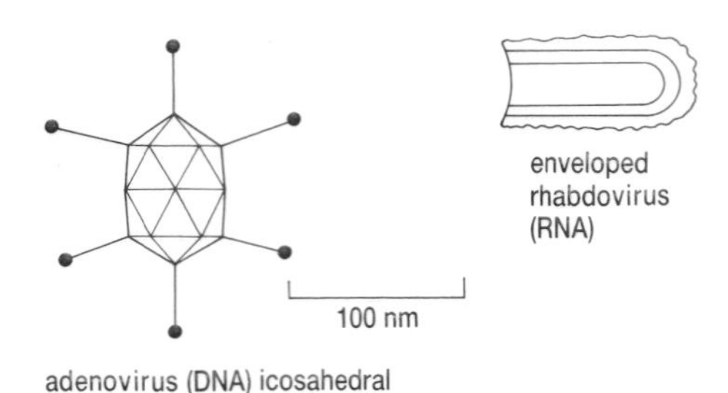

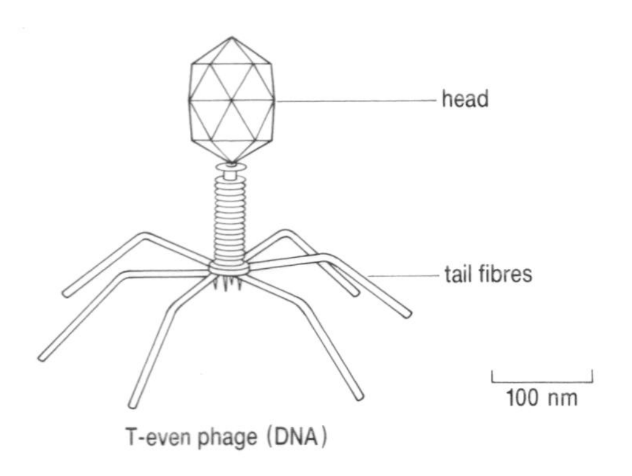

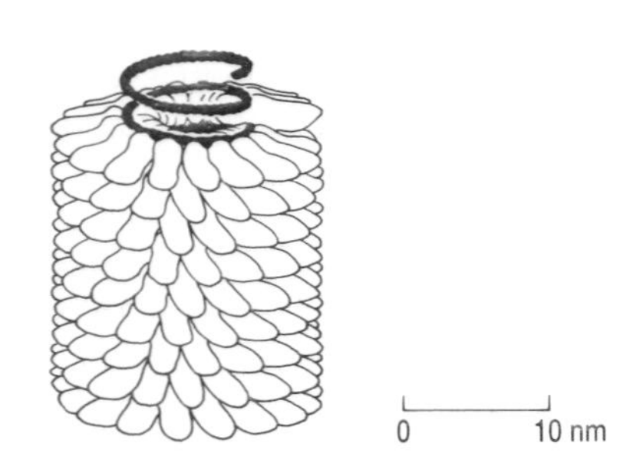

Fig 1.29 Viral structures.

Some bacteriophages, notably the T-series, have a complex structure. These have protein 'tails' which help in infection. The proteins are able to combine with chemicals found in host cell membranes allowing the virus to adsorb onto its host. Once the virus has entered the host cell it quickly uncoats and only nucleic acid can be detected.

Reproduction

The life cycle of any virus is affected by its lack of mobility. They cannot move independently and have to rely on passive dispersal or the use of a vector to transmit them from one host to the next. Most plant viruses have to be introduced into their host plant by the vector and then rely on being transported by fluid streams. Viruses are moved about in fluids randomly and can only infect when they encounter a suitable cell surface.

Reproduction involves the entry of one or more viruses into a cell, multiplication within the cell and then release from it. The pattern of events is often known as the lytic cycle as it may end in cell lysis to release the viruses. The cycle starts with virus entry. Animal viruses adsorb onto chemical receptors in the host cell membrane and are then taken through it. Bacteriophages inject their nucleic acids into bacterial cells leaving the capsule outside. Plant viruses cannot enter tough plant tissue by themselves, unless the cells are damaged. Usually they are injected by **vectors** such as greenfly feeding on plant cells. Once inside they can transfer from cell to cell. Inside the cell, animal and plant viruses are uncoated and they enter an **eclipse** phase when infective particles are difficult to detect.

The viral nucleic acid is activated as it enters the cell. Some of the viruses with single-stranded RNA are 'read off' or **translated** by the host ribosomes directly; others may require a complementary strand of RNA or DNA to be synthesised. Proteins, including those needed for duplicating the nucleic acid, are then synthesised, though some DNA viruses may have replicating enzymes enclosed in the capsid or may be able to use host enzymes for duplicating and transcribing their DNA. Viral nucleic acid molecules have two functions: they act as templates for the synthesis of more nucleic acid for packaging into progeny viruses, and they also carry genes for viral proteins needed for the synthesis and structure of progeny particles. Normal host cell metabolism and DNA activity is often severely disrupted by the activity of the virus.

As viral nucleic acid and proteins accumulate within the cell, capsid proteins take on their three-dimensional forms and the capsid assembles spontaneously. The nucleic acid is packaged, then the viruses leave the cell. Some are liberated by cell membrane rupture, or they may pass through the host cell membrane, picking up an envelope on the way. **Lysogenic** bacteriophages, or temperate viruses, are bacteriophages which, when they invade the cell, do not undergo the full infective cycle but are found only as DNA and are known as **prophages**. They are duplicated with host DNA and most of their genes are repressed. Lambda (λ) phage integrates into the host chromosome; others may exist as plasmids. These viruses can be induced to replicate by stresses such as ultraviolet light or by biochemical changes in the cell. In some bacteria, for example the diphtheria bacteria, the presence of certain prophages enables the bacteria to produce a toxin which they could not otherwise make.

Viral infections are usually difficult to control and treat because any agent that interferes with virus nucleic acid or protein synthesis is almost certain to have an equally harmful effect on host cells. Our knowledge of viruses has given us insight into fundamental genetic processes occurring in cells, and given us some of the techniques used in genetic engineering.

The **Retroviruses** are RNA viruses that carry an unusual enzyme in their capsules called **reverse transcriptase** which can synthesise a strand of DNA from the strand of RNA. They can then synthesise a complementary strand of DNA to form a circular chromosome. This chromosome migrates to the nucleus of the host cell and is inserted into the host cell DNA, thus transforming the host cell. It is then used for the synthesis of viral proteins like any other length of host genome.

This group of viruses has become well known as one of its members, human T-cell leukaemia virus strain III, or HTLV-III, has been established as a cause of acquired immunodeficiency syndrome, usually referred to as AIDS. Rous sarcoma virus is another member of the group which causes a form of cancer in chickens; one of the few viruses linked directly with cancer.

QUESTIONS

1.28 Describe, possibly by a flow chart, the lytic cycle of a bacteriophage.

1.29 What is the function of the tail fibres of a bacteriophage?

1.30 Is the antibiotic penicillin useful in treating virus infections?

1.31 What is the difference between a capsomere and a capsid?

1.32 Describe the structure of a named DNA virus.

1.33 How can micro-organisms survive unfavourable conditions?

1.34 Describe the role of micro-organisms in the cycling of matter through ecosystems.

1.35 Compare reproduction processes in micro-organisms with those of mammals.

SUMMARY

Cells are divided into two types: prokaryotes which lack internal compartmentalisation, and eukaryotes which have sub-cellular organelles and a more complex metabolism. Bacteria and blue-green bacteria are prokaryotic; fungi, protozoa and green algae are eukaryotic. Viruses do not fit into either category as they do not have a cellular structure.

Micro-organisms are found everywhere, reproducing rapidly by binary fission or by spores to colonise every ecological niche. Blue-green bacteria and green algae are the most important carbon fixers in aquatic ecosystems and a wide variety of other habitats. Blue-green bacteria are important nitrogen fixers in nutrient-poor habitats. Protozoa are fully independent eukaryotic unicellular animals. Most are detritus feeders or predate other micro-organisms and are important in decay processes. Euglenoids are flagellate photoautotrophs lacking a cell wall but are heterotrophic in the dark. Fungi are multicellular heterotrophs and many can degrade materials that other micro-organisms cannot; they are among the initiators of the decay process. Viruses are parasites whose activity within cells usually disturbs host cell metabolism. Bacterial respiration may not be oxygen-dependent and they can colonise oxygen-poor ecological niches. Some bacteria gain energy from inorganic chemical reactions and are important geochemical agents.

A table summarising the other groups of micro-organisms is given on page 32.

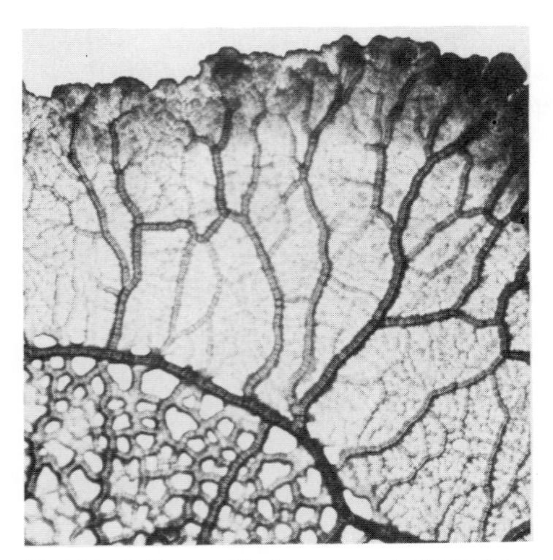

Fig 1.30 A slime mould colony can reach a large size and take on very complex shapes.

Table 1.4 Other groups of micro-organisms.

The Mollicutes.
These are a sub-group of bacteria including the mycoplasmas. They are like bacteria but they lack a rigid cell wall so they are flexible and can squeeze through small gaps. They are insensitive to many antibiotics. They are all parasitic fermenters with complex growth requirements. They are found on animal mucous and synovial membranes where they cause many disorders. Others cause diseases in insects and plants. They are a nuisance in cultures of human or animal cells as they cannot easily be controlled using standard antibiotics.

The Chlamydias.
These are obligate intracellular bacterial parasites, but they form a resting dispersing stage. They do not seem to have any ATP-generating systems of their own and are totally reliant on the host cell for their energy needs. One group causes psittacosis, a disease of birds that can be transmitted to the handlers causing lung disease. The other group causes a variety of diseases such as trachoma, a major cause of blindness in underdeveloped countries, and infections of the urino-genital tract which are difficult to diagnose.

Viroids.
These are nucleic acid molecules, usually RNA, without a protein coat that can infect cells. They are implicated in several plant infections, such as potato spindle tuber, but the mechanisms by which they damage host cells are unknown. They are particularly resistant nucleic acids, heat- and UV-stable. It has been said that some viroid genomes are similar to some of the non-coding portions of eukaryotic genes which are spliced out before mRNA is released into the cytoplasm. It has been suggested that they may interfere with the normal cell mRNA editing process.

The Rickettsias.
These are obligate intracellular parasites whose exact nature is not understood though they were first discovered in 1909. They appear to have a bacterial structure which can survive intracellular digestion. Most are animal parasites transferred by lice or ticks; humans are infected accidentally. They cause diseases such as typhus and Rocky Mountain spotted fever. Epidemic typhus is a specific disease of humans as it is spread by the human body louse which leaves infected faeces on the skin. These get into the body when the louse bite is scratched.

The slime moulds.
These have often been grouped with fungi as they can form a mass of multinucleate cells. However the cells have no rigid walls and the mass of cells is mobile, flowing in an amoeboid motion taking in particulate food. A typical colony is fan-shaped with a spreading leading edge which may be large. In dry substrates the slime mould may form a complex fruiting body releasing spores. Flagellate gametes develop from these which fuse in pairs and form a new slime mould. One sub-group is unique as they exist as uninucleate cells similar to protozoa which multiply by binary fission, but can aggregate and cooperate to make a fruiting body producing asexual spores, yet each cell never loses its individuality.

Satellite viruses.
These are small pieces of nucleic acid that need the aid of another virus to replicate. Many are very small RNA molecules which need proteins from a helper virus or host cell to complete their replication and subsequent infection of new cells. The activity of the helper virus may be slowed down, but satellites can also make helper virus infections worse. There are reports of infectious proteins called prions causing diseases, but little is known about these.

Chapter 2

GROWING MICRO-ORGANISMS

LEARNING OBJECTIVES

After studying this chapter you should be able to:

1. describe how physical factors in the environment limit the growth of micro-organisms and how they may be adapted for extremes;
2. explain how pure cultures of micro-organisms can be isolated from a mixture of species;
3. describe the ways in which micro-organisms and cells are cultured;
4. outline the ways in which micro-organisms can be counted and their growth measured;
5. explain the differing patterns of growth of different kinds of organisms;
6. describe the methods used to grow plant and animal cells in culture.

2.1 MICRO-ORGANISMS IN THE BIOSPHERE

Micro-organisms abound in the oceans, freshwater, the lower atmosphere and on land surfaces. They have been recovered from the depths of the Pacific Ocean where it is cold, dark and under high pressure and from the stratosphere where there is little air, high UV levels and few nutrients; all ecological niches contain micro-organisms of some sort. Some microbial activity is powered by sunlight using photosynthetic processes, other micro-organisms use chemical oxidations and geothermal energy to drive the transfer of energy and materials through ecosystems.

Micro-organisms are easily dispersed and capable of rapid exploitation of a favourable niche.

Micro-organisms can spread everywhere as they are small, light and very easily dispersed by wind and water. Wherever conditions are suitable they grow rapidly and use materials from their surroundings. As they grow they excrete metabolites and gradually change their environment. The changes may make the niche more unfavourable for their own growth, slowing it down, but more favourable for the growth of another species instead, which flourishes. Very few individuals survive unfavourable conditions, but those that do grow rapidly when conditions become favourable again.

Micro-organisms can quickly exploit an environment in favourable conditions because they have a rapid metabolic rate. Their large surface area to volume ratio allows them to exchange materials with their environment very quickly. They can take in large amounts of nutrients and easily dispose of waste materials. They produce a wide range of enzymes and can metabolise a variety of substrates. When conditions are favourable they multiply quickly; bacteria may reproduce every 20 or 25 minutes.

Some ecological niches are favourable for the growth of micro-organisms, for example animal mucous membranes are warm, moist and

Fig 2.1 The very pale streaks across the snow are caused by red algae – another organism which can survive harsh environments.

have a range of nutrients available but there are also harsh environments with several extreme environmental factors. Table 2.1 illustrates these. Arctic waters, a frozen chicken and the stratosphere are all are low temperature environments but each poses different problems for the organisms growing in them. Micro-organisms have evolved strategies allowing them to grow in extreme habitats, and, in even more harsh situations many can survive though not grow. The ability to survive harsh environments is not unique to micro-organisms, but they are far better at coping in these circumstances than other sorts of organisms.

Not every species of micro-organism grows in every ecological niche – far from it. Some species may be capable of tolerating a wide range of environmental factors, but others will be specialised for a very narrow range of environments or may be able to exploit just a few substrates.

The growth of animal and plant cells is also included in this chapter. These can be isolated and cultivated in similar ways to micro-organisms in order to obtain their products.

Table 2.1
Some difficult ecological areas

Area	Problems
hot springs	high temperature high sulphide concentrations
mountains	cold high UV light intensity few nutrients little free water
streams running from peat bogs	acid pH high tannin concentrations
peat bogs	acidic pH low oxygen levels
carnivorous animal dung	high nutrient concentrations alkaline pH high temperatures low oxygen concentrations

2.2 FACTORS AFFECTING GROWTH

Temperature

Temperature is one of the most important environmental factors affecting the growth of micro-organisms. Most species have a characteristic range of temperatures in which they can grow, but they don't grow at the same rate over the whole of their range. Microbial growth is governed by the rate of chemical reactions catalysed by enzymes within cells. An increase in temperature leads to an increase in the rate of reactions; more materials are synthesised and growth is faster. There is, however, an upper temperature limit for growth because heat affects the stability of enzymes and structural proteins. Proteins start to lose their three-dimensional shape at temperatures over 45°C, their properties are altered and enzymes lose their activity. At 60°C over half the proteins present in the cytoplasm of a non-adapted bacterium will have been denatured. Growth stops very quickly when the temperature is high enough to denature proteins. At the lower end of the temperature range we would expect growth to be slower and slower,

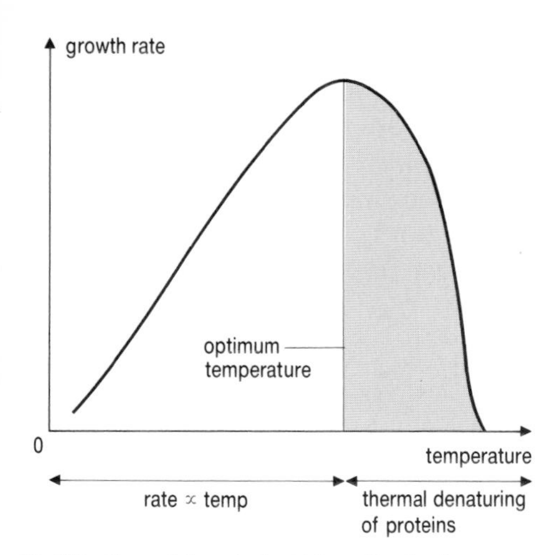

Fig 2.2 The variation of microbial growth rate with temperature.

eventually ceasing when the cytoplasm freezes, but growth of most micro-organisms stops well before this. At low temperatures protein structure changes again and lipids in cell membranes change their physical nature and membrane functions are disturbed. Fig 2.2 illustrates the rate of growth of a typical bacterium over its temperature range. Notice that the lethal temperature is only a few degrees higher than the temperature at which the micro-organism grows most quickly, its **optimum temperature**.

Though vegetative cells are killed quickly at 60°C, some bacterial spores can endure temperatures over 100°C. Many micro-organisms are adapted to survive at high temperatures but few can tolerate sudden changes in temperature. A sudden drop in temperature of 20°C will kill a proportion of the population.

Micro-organisms are put into three overlapping groups according to their optimum temperature and range though not enough is known about some species to allocate them to a particular group (see Table 2.2.).

Table 2.2 Temperature range of some important bacteria.

Organism	Temperatue range for growth (°C)	Significance
Psychrophiles		
Bacillus globisporus	-10 – +20	food spoilage organisms
Micrococcus cryophilus	-10 – +20	in deep freeze
Mesophiles		
Escherichia coli	5 – 40	animal gut inhabitant
Lactobacillus delbruckii	17 – 50	found in milk
Neisseria gonorrhoeae	30 – 40	human pathogen
Thermophiles		
Sulpholobus	55 – 85	hot acid springs inhabitant
Bacillus stearothermophilus	35 – 70	found in milk and cheeses
Bacillus coagulans	25 – 65	pathogen

Psychrophiles

Psychrophiles will grow at 0°C and may have optima below 20°C – they may have proteins and lipids which are cold stable. About 80 per cent of the earth's surface has a temperature of 5°C or below, including much oceanic water, the poles and most of the atmosphere. There are psychrophiles in most groups of micro-organisms and they can be a serious problem in frozen food storage.

Mesophiles

These are micro-organisms which grow in the mid-range of ambient temperatures and have optima between 20 and 40°C. Most of the common species fall into this category, including bacteria, fungi, yeasts, viruses, animal and plant cells. Some human pathogens have a very narrow temperature range for growth.

Thermophiles

Thermophiles grow at high temperatures with optima over 45°C. Organisms which are adapted for life at high temperatures usually have heat-stable proteins. Hot ecological niches include sun-heated rocks and soils, hot springs and power station cooling towers. A temperature of 60°C may be generated by microbial activity in compost, slag and coal heaps, animal dung and hay. Heterotrophic bacteria are the most tolerant, but photosynthesisers are limited to niches below 70°C. Eukaryotic organisms such as fungi and protozoa have lower growth limits than bacteria.

Oxygen concentration

The effect of variation in oxygen concentration depends on an organism's metabolism. Aerobic organisms such as fungi, protozoa and many bacteria use oxygen for respiration, and if the concentrations fall then their activity declines. However some, such as gut inhabitants, need only 2 to 10 per cent oxygen for respiration and are not inhibited until oxygen declines to very low levels. **Microaerophiles** use oxygen but cannot tolerate concentrations as high as 20 per cent and are inhibited at atmospheric concentrations. **Facultative anaerobes** can use oxygen if it is available but ferment if it is not, so their activity is not severely affected by changes. **Obligate anaerobes** do not use oxygen for respiration; aerotolerant species are not affected by the presence of oxygen but other obligate anaerobes are very sensitive to oxygen. Even very small quantities may be toxic as they lack the necessary enzymes to detoxify peroxides and other reactive oxygen groups made in aerobic reactions.

pH

Most organisms are sensitive to changes of pH in their environment. This is because the presence of H^+ ions from acids affects the structure of proteins and other molecules. Most of the natural environment has a pH close to neutral but there are high and low pH environments. Bogs, pine soils, some streams and mine waters are acidic whereas desert soils, animal excreta, decaying protein and some lakes are alkaline environments. Most bacteria grow best at pH 6.5 but can tolerate a range of pH 4–9. **Acidophiles** such as *Thiobacillus thiooxidans* will grow in extreme sites where the pH is 3 or less. Many of these need high H^+ levels to maintain their structures. Nitrate reducers and sulphate reducers are more tolerant of high pH. In general, fungi tolerate slightly more acidic environments than most bacteria, and animal cells are very sensitive to pH changes.

External concentration of solutes and water

The rigid cell wall of a bacterium resists osmotic changes so many bacteria can tolerate a wide range of external solute concentrations. High osmolarities are often due to high salt concentrations, as in salt lakes. Organisms adapted to these conditions are known as **halophiles**. The bacteria *Pediococcus* and *Halobacterium* are halophiles which can tolerate over 25 per cent salt, much higher than that found in the sea! Fungi are very tolerant of high osmolarity, for example they can grow in jam which is over 50 per cent sugar.

Internal solute concentrations of most micro-organisms are higher than the external environment, but if the internal concentrations fall or the external concentration becomes very high, the cells may dehydrate. Many micro-organisms can survive desiccation by forming dormant spores or cysts, but very few organisms can grow where there is little free water available.

Pressure

The ocean floor is a high pressure ecosystem. It is also cold and dark. The average pressure there is about 38 500 Pa which would place enormous stresses on any organisms adapted for surface level. Some organisms do survive and grow, though slowly, in the deeps and many are adapted for life in the gut of deep-sea fish. Some may be **obligate barophiles** which cannot survive at atmospheric pressure. The growth of deep-sea bacteria seems to be too slow to cause problems with deep sea drilling rigs or exploratory equipment. Many micro-organisms make spores which can withstand high pressures and are a problem in materials sterilised using high pressure methods.

QUESTIONS

2.1 In what sort of natural environment would you expect to find
(a) a barophile,
(b) a halophile,
(c) a thermophile,
(d) an acidophile?

2.2 What constituent of bacterial cells is most affected by changes in temperature? How are bacterial cells modified for life in high and low temperatures?

2.3 STERILE TECHNIQUE

For the scientific use of micro-organisms, **pure cultures** are needed, that is cultures of particular species without any other sort of organism in the sample. If other organisms get into the sample it is **contaminated**. Any work done on, or with, that culture would be unacceptable as the results could be due to the contaminating organism. Contamination in an industrial process may mean a whole batch of product has to be thrown away. Systematic precautions are therefore taken to keep cultures pure, or **axenic**, and to stop contamination. The techniques used to handle cultures are called **aseptic techniques**. See Fig 2.3.

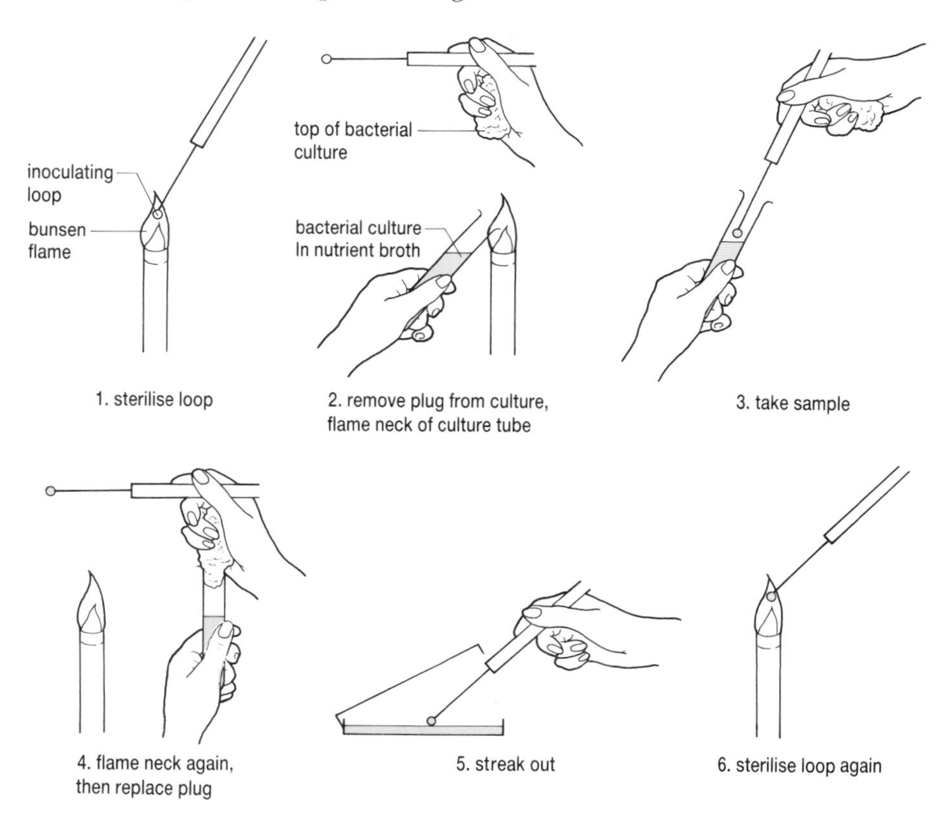

Fig 2.3 Sterile technique.

Pure cultures of micro-organisms contain only one sort of organism.

The same procedures are also used to stop the organisms escaping from their containers and contaminating the workplace or the people handling them. This is particularly important if the organism is disease causing. Everything, including equipment, instruments and materials, must be sterilised before and after use. This is often done by autoclaving but some items, such as culture media, may be affected by heat so other methods have to be used. Solutions can be passed through fine filters which trap micro-organisms. Working areas, some equipment and whole rooms can be sterilised by exposure to ultraviolet light.

AUTOCLAVING

All items used to grow micro-organisms must be free of contaminating organisms.

Autoclaving is the process of heating to high temperatures in a container filled with steam under pressure. This gives a temperature equivalent to 120°C. A domestic pressure cooker works in the same way. The temperature is high enough to kill vegetative cells and bacterial spores.

Work surfaces are wiped with a disinfectant before and after work and contaminated equipment is safely disposed of into a disinfectant. The necks of containers are flamed when they are opened or closed and lids are not left on benches or upside-down to stop air-borne organisms entering them. Anything sterile is not allowed to come into contact with unsterile materials. Particular care has to be taken with liquids containing micro-organisms. A droplet on something which is heated quickly may form an **aerosol** which spreads organisms into the air.

2.4 GROWING MICRO-ORGANISMS

Micro-organisms are grown in mixtures of nutrients called **culture media** in containers. In the laboratory we use test tubes, flasks, bottles or petri dishes but in an industrial process large, stainless steel tanks called fermenters are used. Fig 2.4 shows a pilot-scale fermenter which would be used when testing a method for growing organisms on a large scale before building a production plant. The media and culture vessels must be sterile before they are used. The term ***in vitro*** growth is the growth of micro-organisms or cells in containers. In some instances it may be necessary to grow a micro-organism in its natural host or other living thing, particularly if the organism is disease causing, is parasitic or infects plants. Growth is then described as being ***in vivo***. Micro-organisms can now be grown in controlled environment chambers with all factors carefully regulated.

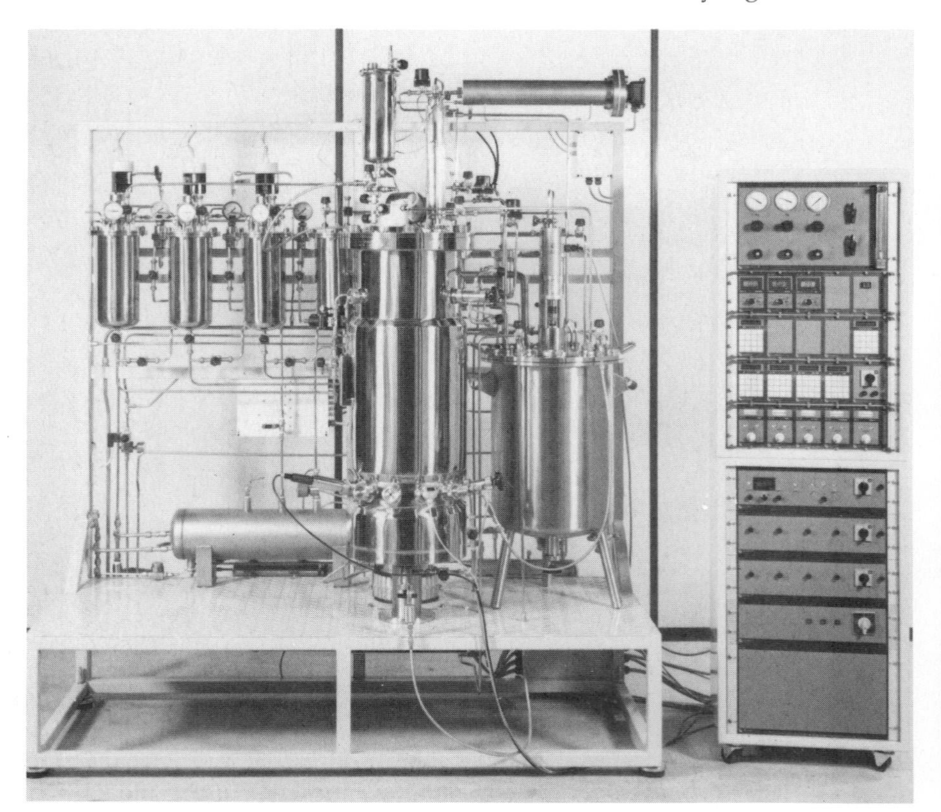

Fig 2.4 Pilot-scale fermenters, with 50 to 200 litres of culture medium, are used to simulate growth conditions in industrial fermentations. The physical and chemical environment is carefully monitored using sensors and is regulated to keep conditions within narrow limits.

Culture media

Micro-organisms need nutrients in the right proportions. Over the years many culture media have been developed, either by trial and error or from knowledge of a species' nutritional needs. All organisms have certain needs in common and any medium has to fulfil them in some way. To start with, all organisms need a **nitrogen source**, which could be nitrogen from the atmosphere, or supplied as inorganic ions, or as complex organic molecules such as polypeptides. Similarly, a **carbon source** is needed such as carbon dioxide or organic carbon compounds like sugars or alcohols. Autotrophic organisms also need an **energy source**, that is light or inorganic compounds. In addition there may be specific **growth factors** which are needed such as vitamins or fatty acids, and a **buffer** is usually added to regulate pH.

All media are based on a buffered mineral solution which is a solution of salts that contains no organic carbon or nitrogen compounds. Few organisms can grow in a solution like this, only nitrogen-fixing photo-autotrophs. If more nutrients are added more organisms can grow. Some materials, particularly minerals, are needed in such small quantities that impurities in laboratory chemicals may be enough for an organism's needs.

General-purpose media

Nutrient agar and similar media, are known as general purpose or **broad spectrum** media as many organisms can be cultivated on them. They provide a range of nutrients which most organisms can use. They are particularly useful for growing organisms whose needs are still unknown. The media contain meat or yeast extract which is a mixture of soluble nutrients including sugars, peptides, amino acids, vitamins, bases and a large selection of inorganic ions. Peptone may be an ingredient, which is partly hydrolysed protein containing phosphates and sulphur as well as organic nitrogen and carbon compounds.

Broad spectrum media contain a wide variety of nutrients in order to grow a wide variety of organisms.

MEDIA

There are many different sorts of media available. Few laboratories make their own – most buy ready-prepared dehydrated media.

This synthetic medium is used to culture fungi:

Czapek Dox Medium (modified by Oxoid)

sodium nitrate	20.0 g
potassium chloride	0.5 g
magnesium glycerophosphate	0.5 g
iron II sulphate	0.01 g
potassium sulphate	0.35 g
sucrose	30.0 g
agar	12.0 g
distilled water to 1 dm^3	

Medium for *Thiobacillus thiooxidans:*

ammonium sulphate	0.2 g
magnesium sulphate	0.5 g
potassium dihydrogen phosphate	3.0 g
calcium chloride	0.25 g
powdered sulphur	10.0 g
distilled water to 1 dm^3	

Thiobacillus obtains energy by oxidising elemental sulphur; few other organisms are able to grow in this medium.

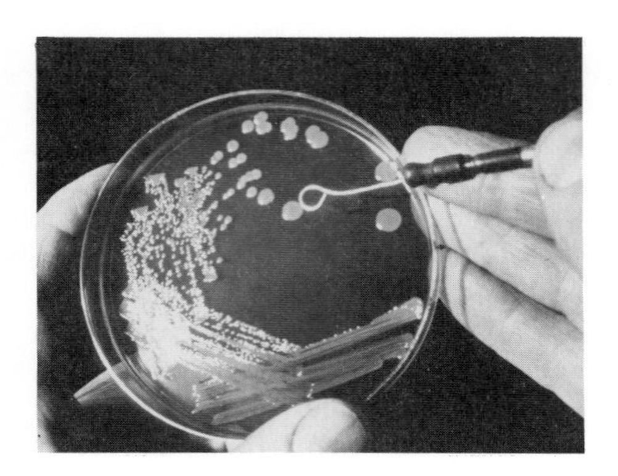

Fig 2.5 Typhoid bacteria, *Salmonella typhi*, grown on a culture medium solidified with agar. The scientist has a technique, described in section 2.5, to obtain single colonies of bacteria and is picking out a sample of bacteria from a single colony using a sterile inoculating loop.

Synthetic media

Synthetic media are made from laboratory chemicals in carefully measured amounts. They may be so specialised that only one or two species can grow on them. If a mixture of bacteria is inoculated into one, only those which can use that particular combination of nutrients will grow; the others will not. These specialised media are called **selective** or **narrow spectrum** media, such as that used for growing *Thiobacillus thiooxidans* shown in the box on the previous page.

A **minimal** medium is one that provides only for the specific needs of a single species with no other factors. Minimal media are useful in investigating biochemical pathways. **Enrichment** media are synthetic media supplemented with materials that promote the growth of a particular species.

These media may include dyes and indicators to help identification, and inhibitors to restrict the growth of undesirable species.

Agar

Media are made up as solutions but they can be solidified if a solid medium is needed. Solutions such as nutrient broth are useful for growing large numbers of organisms. A small inoculum of organisms is added to the medium where they grow as isolated cells making the solution appear cloudy or **turbid**. Adding a gelling agent such as agar, a polysaccharide extracted from seaweed, to a solution will make it solidify.

Agar is useful because very few organisms can metabolise it and it remains solid as micro-organisms grow. Agar is dissolved in water with the nutrients, then the medium is autoclaved and poured into petri dishes. When the medium cools to below 40°C it sets to make a porous semi-solid surface which micro-organisms grow on. Solutes and extracellular enzymes are able to diffuse through it easily. When bacteria are spread over the surface they reproduce to form clumps of cells called **colonies**, which can be seen in Fig 2.5.

QUESTIONS

2.3 Distinguish between a broad spectrum and a narrow spectrum medium.

2.4 Why is it necessary to autoclave media before use?

2.5 What is meant by growing an organism *'in vitro'*?

2.6 Read the two recipes for media in the box on the previous page. Identify in each the nitrogen source and the carbon source. What is the function of potassium dihydrogen phosphate in the second recipe?

2.5 ISOLATING A SPECIES OF MICRO-ORGANISM

A species of micro-organism in 'the wild' usually grows among other species with broadly similar needs, so any sample will contain a variety of species. There are techniques which allow us to isolate a single species of bacterium from a mixture; the same methods are used to isolate fungi. The process depends on bacteria or fungal spores forming colonies on solid media. When a sample of bacteria is inoculated onto a nutrient agar plate, each bacterium multiplies and forms a colony. Even motile bacteria will not travel very far in the moisture film before colony growth can be seen. As long as the initial inoculum contained only a small number of organisms, these colonies will be well spread out, and it can be assumed that each colony is the progeny of a single bacterium. A colony can then be used as a source of a single species of bacterium. Fig 2.6 shows the steps used to obtain a pure culture.

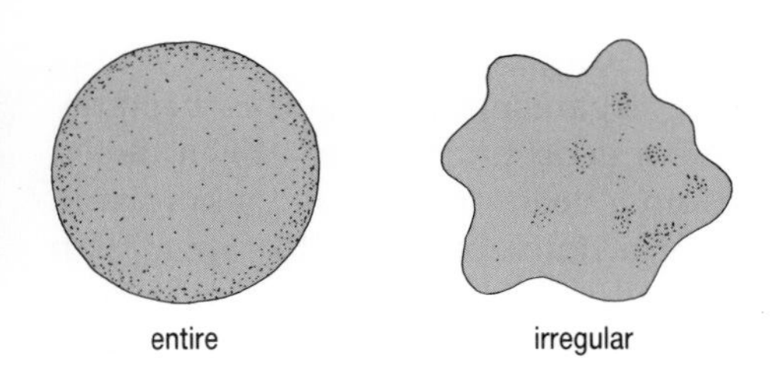

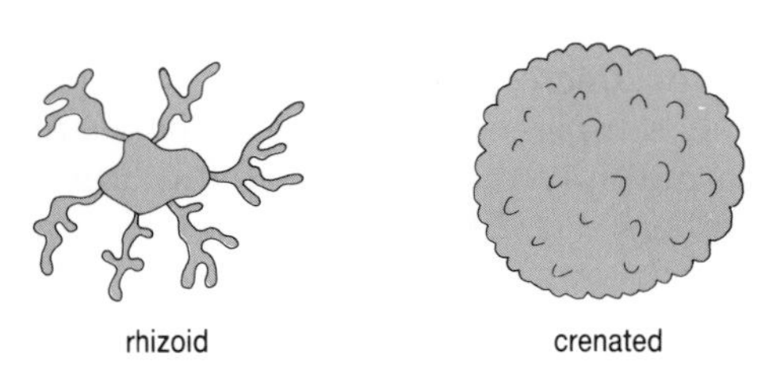

Fig 2.7 Some shapes of bacterial colonies.

The process of taking a sample from a culture and growing it on fresh medium is called **subculturing**. Sometimes a species can be identified at this stage simply from the type of culture medium used and the appearance of the colony. Some species, for example, form colonies with very distinct colours, shapes and textures – some examples are shown in Fig 2.7. Other species may need biochemical tests to identify them. Isolating bacteria is made easier by using a selective medium for the first culture which restricts the number of species that grow.

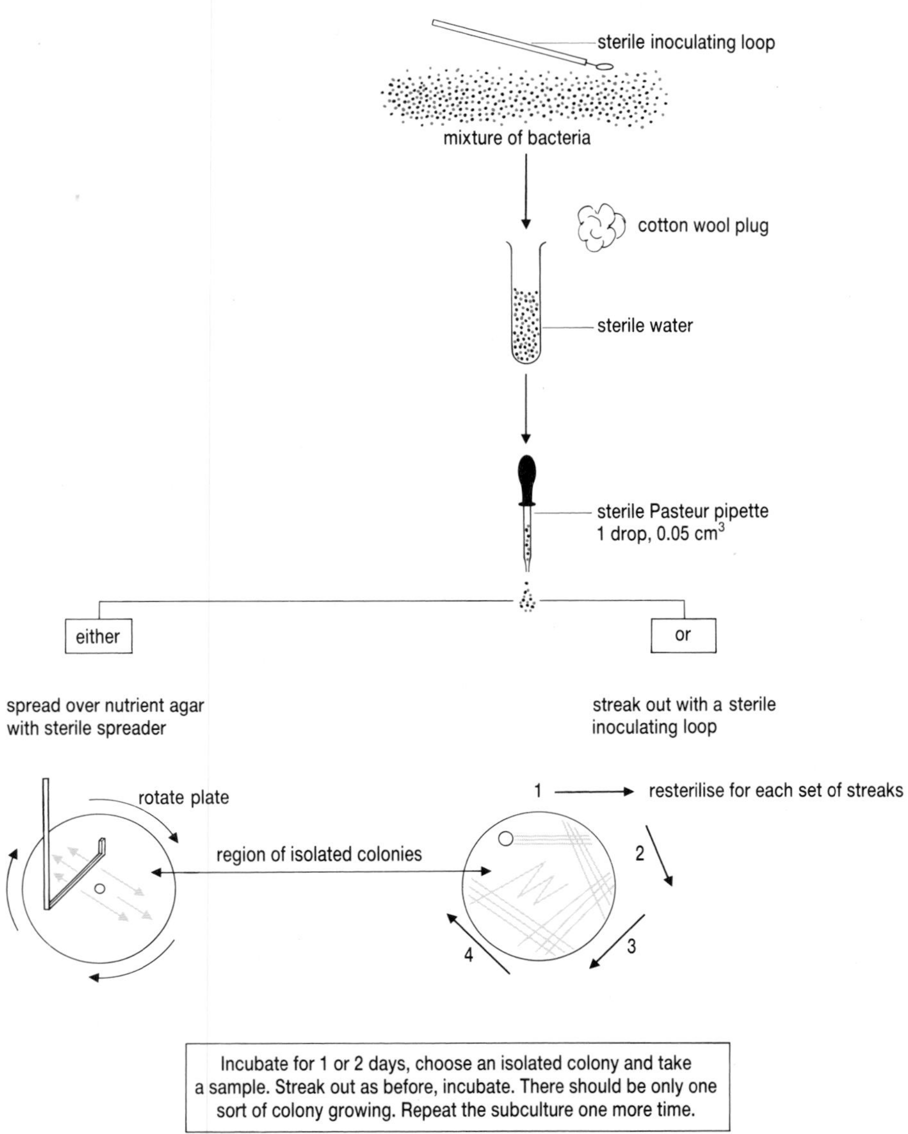

Fig 2.6 Isolating a bacterium to obtain a pure culture.

This method cannot be used for protozoa and algae as they don't form colonies; a dilution method is used instead. The sample is first inoculated into a small amount of sterile liquid medium. This medium, which now contains large numbers of different organisms is then diluted with more medium to make very dilute suspensions containing only one or two organisms per drop. A large number of test tubes of medium are inoculated with a drop of the dilute sample and incubated. Any growth occurring is therefore likely to be from one organism multiplying.

2.6 COUNTING MICRO-ORGANISMS

Large organisms can be counted easily by sampling or direct observation, but this method is not appropriate for organisms as small as bacteria, which are not easy to see except with very good microscopes. Even then a live bacterium can't be distinguished from a dead one just by looking at it. Counting methods often involve the use of microbial growth rather than direct counting. The large numbers of individuals involved also creates difficulties, for example if a sample from a bacterial colony is inoculated into nutrient broth and incubated overnight in optimum conditions, approximately 10^8 organisms will grow per cm^3. A drop with a volume of 0.05 cm^3 will contain about five million cells – far too many to count accurately.

In order to estimate the number of micro-organisms in a sample therefore, it is necessary to dilute the sample until small numbers are reached, then to count, and finally to multiply the count by the dilution factor to arrive at an estimate of numbers. Fig 2.8 shows the process of making successive **serial dilutions** to reach countable numbers.

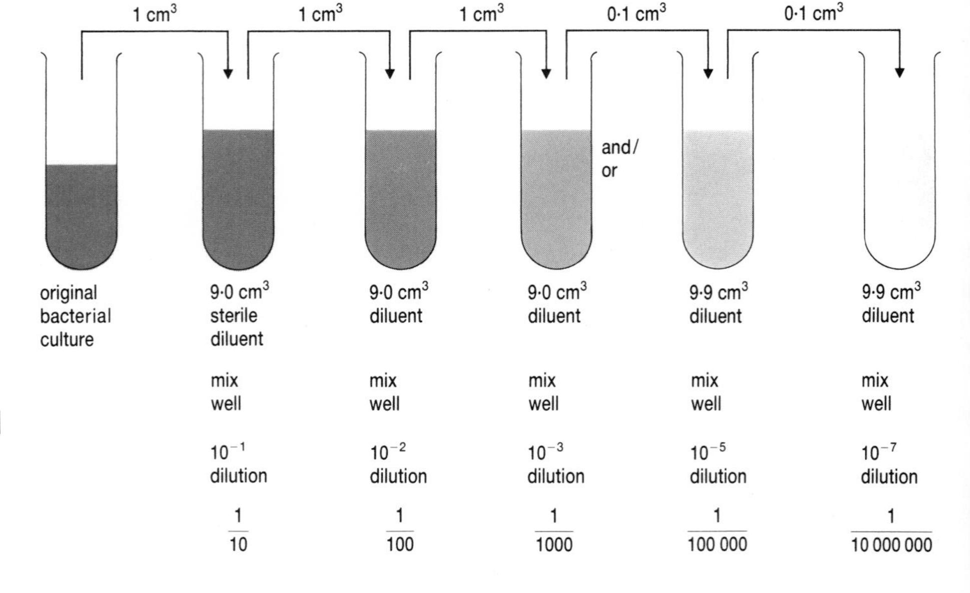

Fig 2.8 Making a serial dilution.

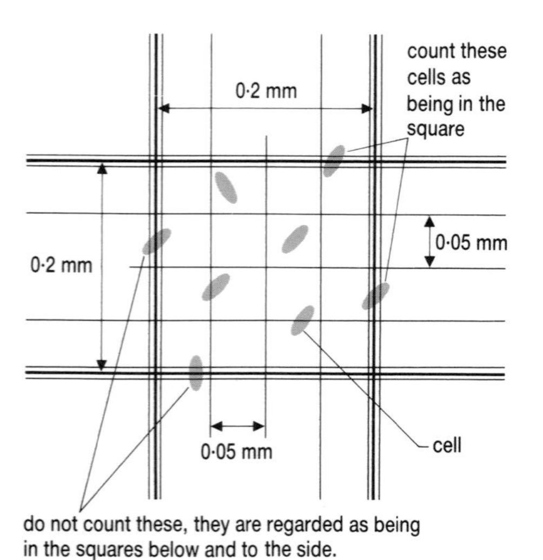

Fig 2.9 **(a)** A haemocytometer
(b) Enlargement of the grid.

Total counts

Direct counts

All the organisms in a small volume of a diluted sample are counted in a suitable counting chamber. A **haemocytometer** is often used, which is a thick slide with a slightly thinner central section, as shown in Fig 2.9. When a coverslip is placed over the central section, a chamber of known depth is created. The central section has a grid etched on it. The size of each square is known and so the volume of liquid over a square can be calculated. Usually it is 0.004 mm^3. A sample of cell culture, suitably diluted, is introduced into the central section and viewed under a microscope. The number of organisms in several squares is counted, an average found and, using the dilution factor, the number per cm^3 in the original sample is estimated.

Fig 2.9(b) illustrates how to count cells with a haemocytometer. Care has to be taken with cells that overlap squares and a procedure is shown in the diagram that prevents the same cell from being counted twice. This method is useful if there are large numbers of cells in the original sample, but it does not tell us how many of the cells are living and how many are dead.

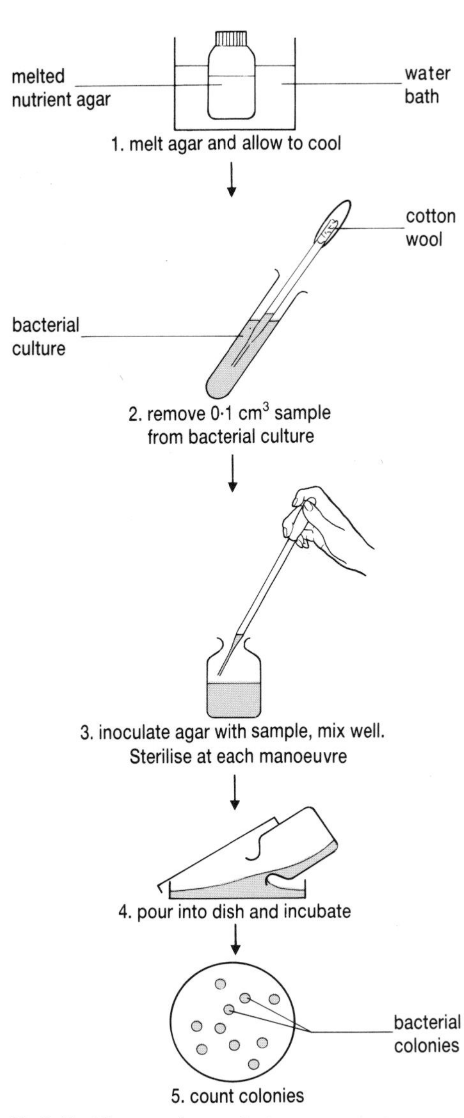

Fig 2.10 The pour plate method to count bacteria.

Viable counts count only organisms or cells that are capable of growing. Direct counts count *all* the organisms present in a culture sample.

Example. In Fig 2.9 there are six organisms in the square, which has an area of 0.04 mm^2. The chamber has a depth of 0.1 mm,
therefore there are 6 bacteria in 0.004 mm^3
and in 1 mm^3 there are 1500 bacteria.
The sample this was taken from was diluted to a 10^{-5} dilution,
therefore there are 1500×10^5 organisms in 1 mm^3 of the original sample
and $1000 \times 1500 \times 10^5$ organisms in 1 cm^3 of original sample.
Expressed in standard form, the number of organisms per $cm^3 = 1.5 \times 10^{10}$ organisms per cm^3.

Indirect counts

These methods depend on the *effects* of organisms to estimate their numbers. As organisms grow they make a nutrient broth turbid. The turbidity can be measured using a **colorimeter** which passes a beam of light through a culture sample and the light absorbed or scattered is recorded. The more organisms there are the greater the optical density of the solution. Another method uses the Coulter counter which was originally devised for counting blood cells. A probe with two electrodes is put into a culture sample. One of the electrodes is inside a small glass tube with a narrow entrance. When a bacterium passes through the entrance it alters the conductivity inside the probe and is recorded. The size of the alteration depends on the size of the bacterium, so a cell size analysis can be done as well. Neither of these methods can distinguish live cells from dead ones, nor can they distinguish cells from particulate matter of a similar size. Their advantage over direct counting methods, however, is that they can both be automated.

Viable counts

Viable counts are those in which only cells capable of growing are counted. A **plate count** is used to count bacteria; the procedure is shown in Fig 2.10. As most samples contain very large numbers of bacteria, they have to be serially diluted before counting. Several dilutions are made and, at each dilution, a measured volume of culture is mixed with melted but cool nutrient agar and poured into a sterile petri dish. Each living bacterium in the suspension will form a colony. Duplicate plates are prepared for each dilution to give a more accurate estimate. The dishes are incubated for two days then the number of colonies counted. This is multiplied by the dilution factor to give an estimate of the number of bacteria in the original sample.

Counting viruses

Viruses do not grow outside cells, so they are counted using a **plaque assay**. Viruses are grown in a culture of animal cells called a **tissue culture**, which is a flat sheet of cells on the bottom of a petri dish; more details of this can be found in section 2.9. When a virus infects a cell, it multiplies inside and its progeny escape by lysing the cell. The neighbouring cells then become infected and lyse in turn, leaving a hole in the sheet of cells which can be seen if the cell sheet is stained. Each hole is called a plaque and it is assumed that each plaque is the result of one virus multiplying. Plaques can be seen in Fig 2.11.

To count viruses in a suspension, serial dilutions of the suspension are made and measured volumes of each dilution are mixed with suitable host cells. The mixture is dispensed into a petri dish and incubated for three days. The sheet of cells is then stained and the number of plaques at each dilution is counted. The number in the original suspension can then be estimated.

Bacteriophages can be counted in a similar way they are mixed with bacteria which are then grown on nutrient agar. The bacteria grow throughout the agar plate except where the virus has lysed them.

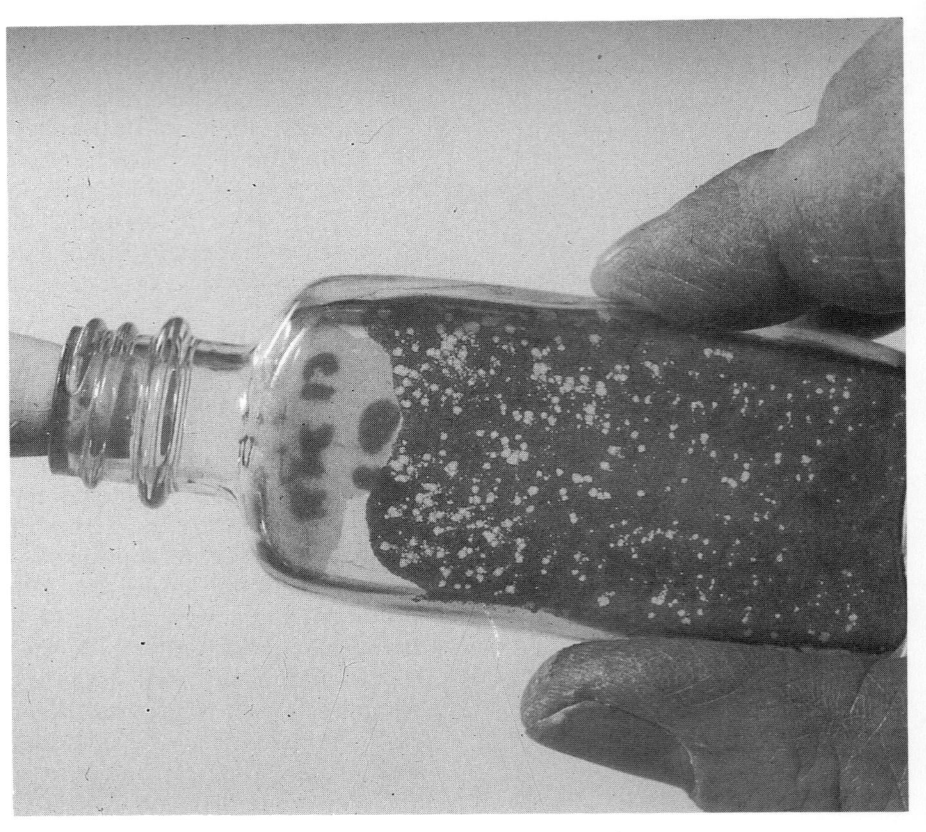

Fig 2.11 A culture of animal cells infected with viruses. The viruses destroy animal cells when they infect, leaving plaques in the cell sheet.

QUESTIONS

2.7 Distinguish between a total count and a viable count.

2.8 If there were 7×10^5 organisms per cm^3 in a suspension, how many organisms would you expect in 1 cm^3 of a 10^{-3} dilution?

2.9 The following counts were obtained from a series of plate counts, each using a 0.1 cm^3 sample per plate, to estimate how many bacteria were present in a water sample.

At a dilution of 10^{-5} there were 87, 73, 91 colonies; at 10^{-6}: 9, 8, 8; and at 10^{-7}: 1, 0, 0.

Work out the average count at each dilution. Use the figure for **each** dilution to calculate how many bacteria were present originally. Can you suggest a reason why they are not the same? Use your figures to estimate how many viable bacteria were in the original culture.

2.7 PATTERNS OF GROWTH

Microbial growth almost always refers to changes in the size of a population not a change in the size or mass of an individual.

Bacteria, protozoa and unicellular yeasts

Most bacteria, protozoa and yeasts reproduce asexually by binary fission or budding. When a bacterium reaches full size, or sooner if conditions are favourable, it will divide into two smaller cells. If the environment provides plenty of nutrients, electron acceptors and growth factors, bacteria grow very quickly and binary fission is rapid. The life span of a cell, called its **generation time**, decreases as conditions improve. In good conditions, the cells do not reach maximum size before they reproduce, so cell volumes are small. When the cells are multiplying at their maximum rate they are said to be in a state of **balanced growth**. All measures of cell growth such as increase in dry weight, DNA content and protein content increase in the same proportion. In every generation there is a doubling of cell components.

A single bacterium will produce two daughter cells when it divides. These in turn produce four cells then eight, sixteen, thirty two and so on, as shown in Fig 2.12. This exponential increase in population size can continue indefinitely as long as the environment is favourable. The population descended from *each* bacterium after a number of generations of balanced growth can be estimated using the general equation:

$$\text{population in the } n^{th} \text{ generation} = 2^n$$

As large numbers are involved, they are usually converted to a logarithm before they are plotted on a graph. In a natural environment micro-organisms cannot multiply exponentially for long before a shortage of an environmental factor necessary for growth or predators reduce the rate of population growth.

Though protozoa and yeasts can reproduce sexually, binary fission is the main method so the pattern of growth of these organisms is very similar to that of bacteria.

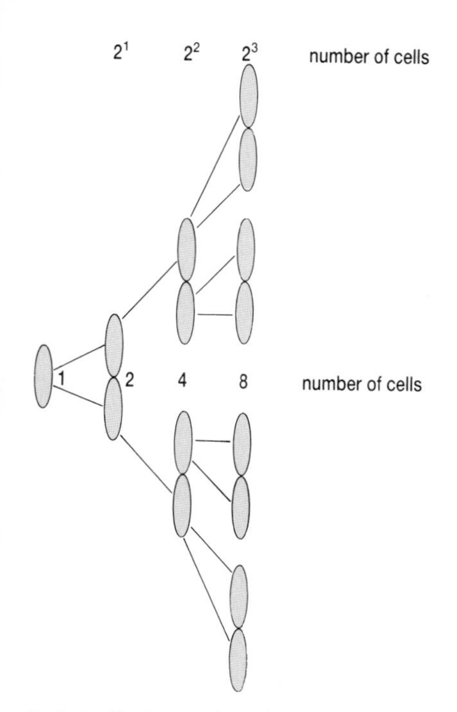

Fig 2.12 The increase in numbers with binary fission.

When cells are in a state of balanced growth they multiply exponentially.

The bacterial growth curve

A bacterial growth curve shows the change in size of a bacterial population with time in a batch culture. To obtain a bacterial growth curve a batch of sterile nutrient broth is inoculated with bacteria. The organisms have a good supply of nutrients, salts, oxygen and other electron acceptors and growth factors. The culture is incubated at optimum temperature and samples are removed regularly for counting. The number of viable bacteria in each sample is counted and the number of organisms plotted as log numbers against time, giving a growth curve like the one shown in Fig 2.13.

As the organisms grow they use materials in the culture medium and they secrete and excrete metabolites into it. As a result the composition of the medium, and hence its favourability, constantly changes. Bacterial growth curves are therefore not exponential, but consist of different phases which reflect changes in metabolic activities through time. The phases are not separate, but change gradually from one to another.

Phase 1 : lag, latent or initial stationary phase

The bacteria are active but there is little increase in numbers. They imbibe water, synthesise ribosomes and induce enzymes to exploit the culture medium. The length of the phase depends on what medium the bacteria grew in before the investigation started, and which growth phase they were in. If the previous medium was very similar and the cells were actively growing, the latent phase is very short. At first, large cells are produced and there is a slow increase in numbers. Gradually the generation time gets shorter, and cells become smaller, as they enter the next phase.

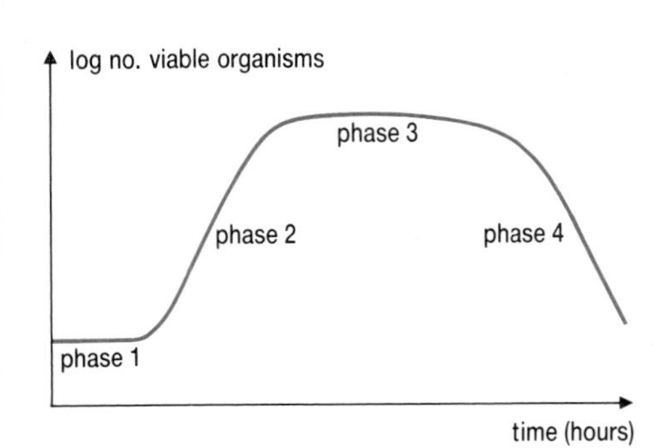

Fig 2.13 The bacterial growth curve.

Phase 2 : exponential or log phase

Cells reproduce at their fastest and the number of bacteria increases directly with time. The generation time is very short but varies with species, for example *E. coli* divides every 21 minutes at its fastest; *B. stearothermophilus* every 9 minutes. The cells are very active metabolically and are in balanced growth. This is the most useful phase for estimating growth rate.

Phase 3 : stationary phase

As the cells grow they alter the culture medium. Nutrients and electron acceptors become depleted and there is a fall in pH as carbon dioxide, acids and other metabolites build up. There are changes in the cells as their energy stores are used up, and the reproductive rate falls. Organisms die in greater numbers leaving only those better able to survive in the difficult conditions.

Phase 4 : final, stationary or death phase

Conditions in the culture medium are very severe and far more bacteria die than are produced so the number of living cells declines.

QUESTIONS

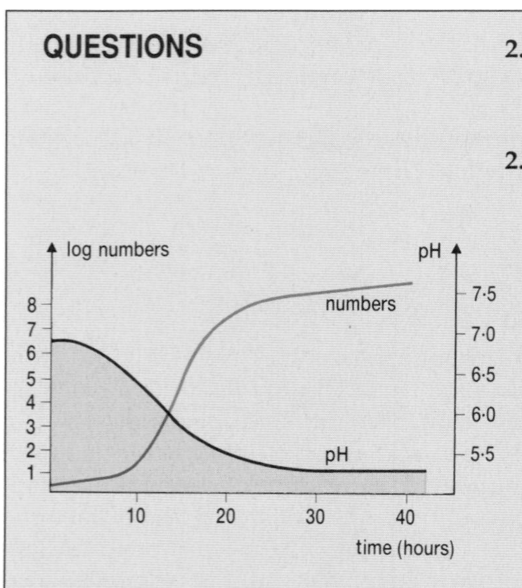

Fig 2.14 Change in pH and cell numbers with time.

2.10 You are provided with a culture of bacteria. Outline a method you could use to determine the number of living cells per cm^3 in the culture.

2.11 The graph in Fig 2.14 shows the growth of a bacterial culture you are investigating. Log numbers of living cells have been plotted against time. The culture medium is a minimal medium with a single carbon source.

(a) Explain the changes in the curve of numbers of living cells between 5 hours and 15 hours, and between 25 and 40 hours.

(b) The pH of the culture medium has also been plotted. Suggest one substance that could have accumulated during the course of the investigation and which brought about the change in pH.

(c) What would happen if another carbon source were added after 36 hours?

(d) Draw a graph of the results you would have expected if the *total* number of organisms had been recorded instead of the numbers of *living* organisms.

Fig 2.15 A 'Fairy Ring' of fruiting bodies of *Marasmius oreades*. Fungal activity causes the breakdown of materials in the soil and releases nutrients, which encourages grass growth too.

Fungi

Mycelial fungi start with the germination of a single spore which produces a single hypha. The hypha branches repeatedly producing a mass of hyphae radiating out in all directions, exploiting the substrate. If there is an even distribution of nutrients, for example on an agar plate a typical circular mycelium develops. Hyphae grow from their tips and the mycelial mat gradually increases in diameter. Eventually the central region is exhausted of its nutrients and the mycelium there dies, leaving a ring of actively growing mycelium. The 'fairy ring' in Fig 2.15 is the above ground part of a large mycelium which has been growing for some time.

The circular mycelial mat is an inconvenient way to grow fungi industrially; liquid cultures are preferred. In a liquid a fungus grows as a floating mass of hyphae at the air-liquid interface. The maximum growth rate, measured as increase in dry mass per unit time, can be achieved when the fungus is grown in a **submerged culture**. In this method, the culture medium is stirred vigorously to keep oxygen concentrations high in the culture vessel and the fungal mycelium breaks up into small pellets. Each pellet has a high surface area to volume ratio which allows efficient gas and nutrient exchange and is capable of exponential growth, as long as it does not grow too large. Once a certain size is reached the centre of the pellet becomes anaerobic and dies which may affect the growth and activity of the rest of the pellet.

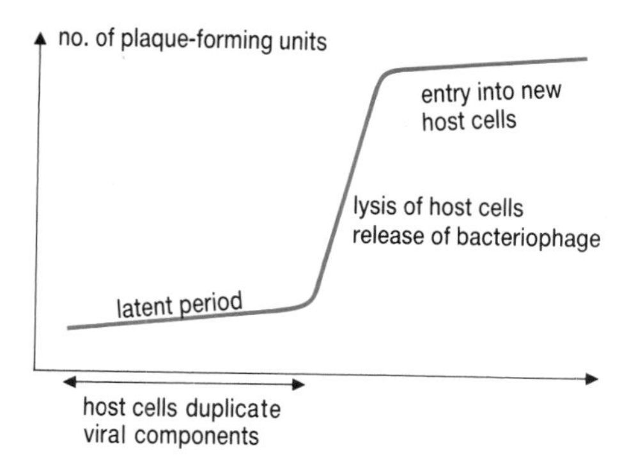

Fig 2.16 Bacteriophage one-step growth curve.

Viruses

Once an individual virus has been assembled it does not increase in size. After it leaves a host cell it adsorbs onto another host cell nearby. Inside the cell, virus replication quickly generates large numbers of infectious virions. When these escape there is a sudden rise in the number of viruses that can be detected. These can then re-infect and repeat the cycle. When virus numbers are measured, they increase in a number of steps corresponding to reproductive cycles. Fig 2.16 shows a typical growth pattern for a bacteriophage, called a **one-step growth curve**.

2.8 CULTURING PLANT CELLS

Plant cells can be grown in culture like micro-organisms. There are several different techniques to choose from depending on the culture's use. They may be grown as suspensions of individual cells, as protoplasts which are plant cells without cell walls, as aggregates of cells called callus culture or as explants of particular tissues.

Callus culture

Entire plants can be regenerated from callus cultures.

A suitable part of a plant or a seedling, such as hypocotyl, is selected and surface sterilised with a disinfectant. A section is cut and this section of tissue, or explant, is transferred aseptically to a sterile container of agar with plant growth regulators. Cells in the explant divide under the influence of the regulators, including many of those that do not normally divide in a whole plant, and a mass of unspecialised cells develops. The cell mass, or callus, takes a few weeks to grow. The callus tissue, shown in Fig 2.17(a), and small samples taken from it, are capable of regenerating a new plant which is genetically identical to the explant source. When many plants are produced from a single callus culture they form a **clone**. A wide variety of plants can be regenerated in this way, but not all the economically important crops have been cloned yet.

(a)

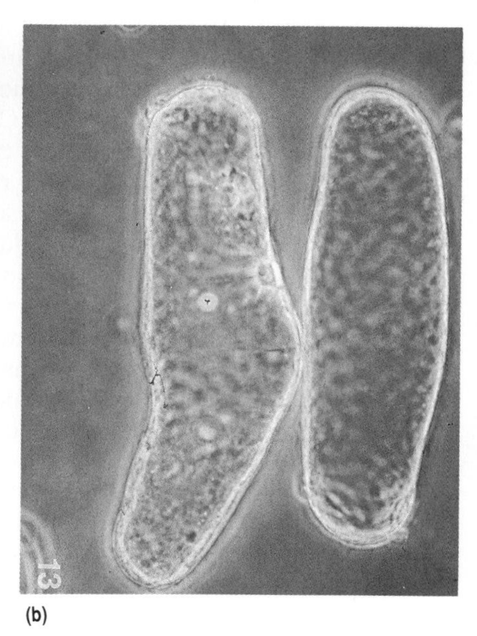

(b)

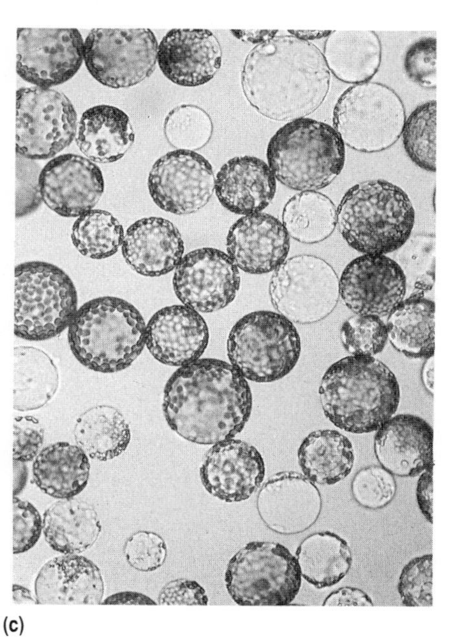

(c)

Fig 2.17 **(a)** Callus tissue of the Madagascar Periwinkle, *Vinca rosea*, cultivated to synthesise anti-cancer drugs. New plantlets grow from the callus tissue.
(b) Carrot cells grown in suspension culture.
(c) Protoplasts from Tobacco, *Nicotiana plumbaginifolia*. The cells become rounded when the cell wall is removed. Most of these cells contain chloroplasts but those which were part of the leaf epidermis are colourless.

Suspension cultures

These are cultures of single cells or small clumps of cells suspended in a suitable culture medium. The culture medium is entirely synthetic and complex; the cells need sucrose (eliminating the need for photosynthesis), plant growth regulators, some amino acids and vitamins. A sample of a callus culture is inoculated into the medium and the cells are dispersed by shaking with small amounts of cell wall-degrading enzymes such as cellulases. The culture has to be aerated, which also helps to break up large clumps. Cultured cells seem to grow best in small clumps rather than as isolated individuals, and it is possible that they in some way condition the medium around them.

The growth of cells in suspension cultures follows the same pattern as that of bacteria, but is much slower, and cell cycles and generation times are longer. A large inoculum of cells is needed to get the culture started, then there is a long lag phase. The exponential phase is short with very few generations before the culture enters the stationary phase.

Suspension culture techniques are being modified for the industrial production of plant metabolites. For example, plant cells are immobilised on beads of an inert material and immersed in culture medium. This is thought to prolong the life of the culture.

Protoplast cultures

To make protoplasts, the cell walls are removed from plant cells by gentle enzyme action, usually a mixture of cellulases and pectinases. Great care is needed in culturing protoplasts if the fragile cells are not to suffer osmotic shock; the water potential of the medium has to be very carefully regulated. Protoplasts, like those shown in Fig 2.17(c), are used for investigations into plant cell activity; to form hybrid cells which are used for genetic investigations; and to generate hybrid plants with new mixtures of characters.

2.9 GROWING ANIMAL CELLS

Many kinds of tissue can be taken from living organisms and persuaded to grow in the laboratory. These are **primary cell lines**; usually they will grow and multiply for a while but eventually the line dies out. Occasionally a variant cell line arises that continues to grow indefinitely and which can then be grown on a large scale. Other permanent cell lines arise from tumours. These are **transformed** cell lines and the cells are relatively unspecialised. It is also possible to maintain whole sections of tissues and organs in culture for a while; these are **tissue** and **organ cultures**. Many of the cell lines commonly in use today are derived from embryonic tissues which still have the capacity to divide frequently.

Few types of animal cells will grow in culture successfully for many generations

To make a cell line the tissue is sampled or dissected out and the cells gently dispersed by enzymic and mechanical action. They are then mixed with a culture medium and dispensed into petri dishes or bottles to settle and grow as a layer just one cell thick, called a **monolayer**. The cells grow as a single layer because they are subject to **contact inhibition**. This means that as they divide new cells move around, but when they come into contact with each other they stop. The cells are harvested by treating them with an enzyme such as trypsin which breaks cytoplasmic links between cells and the substrate.

The culture medium is complex, and even now we cannot make a completely synthetic version. It is usually a solution of salts with glucose and vitamins, though recipes vary slightly. Other factors are needed in small amounts and are supplied by adding about 10 per cent horse or calf serum. There is also an indicator in the medium to give a visual indication of the pH in the culture medium. Additionally, human and animal cells need an environment with a higher proportion of CO_2 in than the atmosphere. Everything has to be done aseptically to prevent contamination; sometimes there is an antibiotic in the medium, for example when using cells to grow viruses, but this isn't always possible.

The pattern of growth of animal cells is similar to the bacterial pattern. A large number of cells is needed to start the culture and there is a long stationary phase. The cell doubling time in the exponential phase is about 15 hours. Though animal cells have long been grown on a small scale, it is only recently that large-scale methods have become widely used. Efforts are being made to develop systems in which the cells are immobilised on beads of inert material, but this method cannot be used to grow the important monoclonal antibody-making cells.

Once the cells are grown they are used in a variety of ways: for basic research; to investigate the effect of drugs and other chemicals on their growth; to grow pathogens in; for vaccine production; or for the production of metabolites such as collagen or monoclonal antibodies. More details about these can be found in Chapter 5. They are even being investigated as a substitute for skin. Reconstructed human skin is a mixture of fibroblasts and collagen which is incubated for a few days to form a mesh before a small sample of a person's epidermis is put in the culture. The human epidermis grows outwards producing a tissue. Although it lacks many of the structures of real skin, it is a very useful research tool and may one day be used for skin grafts.

QUESTIONS

2.12 What are the main steps in establishing a culture of animal cells? How does this differ from setting up an isolated plant cell culture?

2.13 **(a)** How would you use bacterial culture techniques to investigate the activity of a newly discovered antibiotic against a bacterium, *Bacillus megaspottium*, isolated from a local patient?

(b) How would you modify your investigation if you were looking at an anti-viral agent?

(c) How could you investigate whether the antibiotic had toxic effects on human cells?

2.10 BATCH AND CONTINUOUS CULTURE

The technique described in section 2.7 to investigate the bacterial growth curve is a typical **batch** culture process. The organisms or cells are inoculated into a vessel with a supply of materials for growth. They grow until the medium becomes too unfavourable or some factor becomes limiting, then the culture declines. Cells, or any product which the organisms have made, can be harvested from the culture. Several batch processes are described in Chapters 3 and 5.

Batch culture is not always the best way of obtaining cells or a cell product. For example, if a steady supply of cells is needed then a culture of cells in exponential phase is best and conditions have to be continually adjusted so that they do not enter the stationary phase. Similarly, some chemicals are only made by cells in certain growth phases so growth in that particular phase would have to be maintained. **Continuous culture** techniques are used to grow organisms continually in a particular phase. A main culture vessel is used and the state of the culture medium is monitored constantly. If there is a shortage of a nutrient, oxygen or any other ingredient, more is added to bring up levels to the optimum, if the pH changes it can also be adjusted back to optimum.

Cells can be maintained in a particular growth phase in continuous culture.

Fig 2.18 shows a simplified diagram of a continuous culture vessel that could be used in a laboratory. Growing cells and micro-organisms in continuous culture industrially, where vessels may hold thousands of litres, poses even more problems. The large-scale growth of micro-organisms in fermenters is discussed in Chapter 5.

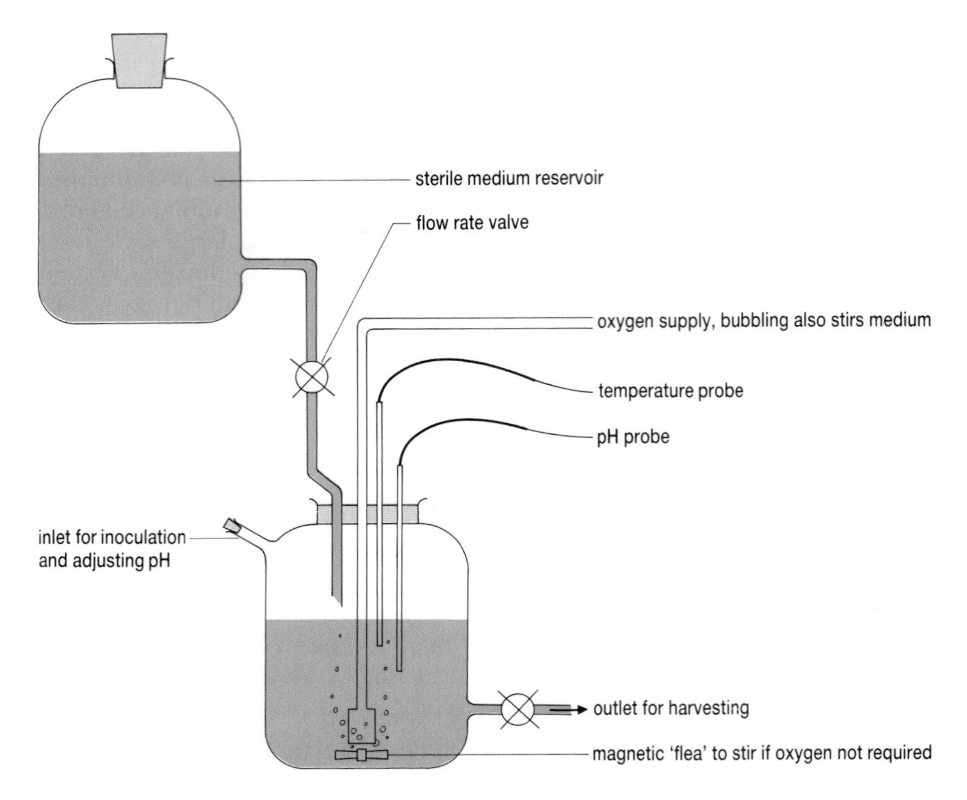

Fig 2.18 A continuous culture system for a laboratory.

SUMMARY

Micro-organisms have a wide distribution and can be found in extreme environments. There are upper and lower temperature limits to microbial growth and there may be biochemical adaptations in cells to extreme environments. Micro-organisms' growth is also affected by pH, external concentration of solutes, oxygen levels and pressure.

The culture of micro-organisms requires aseptic techniques to keep out contaminants. Micro-organisms can be grown in artificial culture media – the precise composition of the medium determines which organisms can grow on it.

Micro-organisms can be counted directly or their numbers estimated by their effects. Some methods of counting distinguish between living and non-living cells.

Micro-organisms have characteristic growth patterns which are limited by changes in the environment.

Cells can be grown in batches, or they can be maintained in a particular growth phase in continuous culture. Plant and animal cells can be cultivated by methods similar to those used for micro-organisms.

Theme 2

ENVIRONMENTAL AND INDUSTRIAL BIOTECHNOLOGY

Mr. Bosher, upon being appealed to for his opinion, explained that science was alright in its way, but unreliable: the things scientists said yesterday they contradicted today, and what they said today they would probably repudiate tomorrow.

Robert Tressell, *The Ragged Trousered Philanthropists*

Mr. Bosher, speaking one hundred years ago, was right, but not in the way he intended. Scientific advances are made daily and it is difficult to keep up with developments; what one thought was true one day has disappeared the next. Mr. Bosher would have been astounded to find out that today a whole forest can be raised in an incubator the size of a domestic fridge-freezer or that bacteria can be made to make human hormones.

New advances may have profound significance, such developments as the large scale manufacture of antibiotics and ingredients for the contraceptive pill by fungi have transformed our lives. Industries using micro-organisms have burgeoned over the last two decades because processes using micro-organisms, or their products, are often cheaper to run or more efficient and many exciting new products have been developed. Advances in our understanding of, and ability to manipulate, a cell's genetic material have allowed us to design cells to carry out specific tasks in ways previously thought impossible.

This section is about the use of cells and micro-organisms to make products or to carry out chemical transformations. These activities include making chemicals and enzymes as well as more traditional products such as food and drink. Decay organisms are harnessed to degrade wastes into useful materials and fuels and to clean up the environment.

Prerequisites

Read sections 1.6, 1.7, 1.12 and sections 2.4, to 2.10 which cover the biology of commercially grown organisms and how cells and micro-organisms are grown. An understanding of GCSE chemistry, the carbon and nitrogen cycles, and enzyme biology is helpful, but no previous knowledge of food science is required.

Chapter 3

THE MICROBIOLOGY OF FOOD

LEARNING OBJECTIVES

After studying this chapter you should be able to:

1. explain why micro-organisms are found on food;
2. describe how micro-organisms can be used to make or flavour foods;
3. describe the spoilage problems caused by micro-organisms;
4. understand the main ways of preserving food;
5. appreciate some of the problems of some preservation techniques.

3.1 MICRO-ORGANISMS AND FOOD

All organisms need materials from the environment. Many micro-organisms have needs similar to our own, and so food supplying our needs is suitable for a wide range of micro-organisms too.

Micro-organisms growing in food secrete extracellular enzymes into it. These degrade complex organic molecules into smaller soluble molecules which are absorbed across the cell wall and membrane. The enzyme action brings about changes in the appearance and texture of the food. New flavours and aromas may develop from breakdown products and from the presence of metabolites. There may even be obvious bacterial colonies or fungal hyphae. Sometimes we want these changes, for example for making milk into cheese or fermenting soya beans to make bean curd (see Fig 3.1). At other times, though, the changes are undesirable: food may become slimy or develop 'off' flavours, unpleasant smells are generated and the

Table 3.1 Typical surface flora of plants and animals

Plants	Animals
Alcaligenes	*Achromobacter*
Flavobacterium	*Coliforms*
Lactobacillus	*Lactobacillus*
Leuconostoc	*Micrococcus*
Micrococcus	*Pseudomonas*
Pseudomonas	*Staphylococcus*
Botrytis	
Monilia	

Fig 3.1 All these foods from a supermarket trolley have been made using micro-organisms, even black tea leaves are fermented before they are dried.

texture breaks down. Some food contaminants can cause disease, for example the toxin-producing fungi and *Salmonella* bacteria which cause food poisoning. Food processing has to be conducted so that it encourages the desirable changes and inhibits the growth of contaminating organisms.

3.2 WHERE DO THE MICRO-ORGANISMS IN FOODS COME FROM?

Fig 3.2 The film of wild yeasts growing on the outside of a grape gives it a greyish 'bloom' covering the shiny grape skin.

Micro-organisms in foods come from a variety of sources. Fresh vegetables carry soil organisms from the fields; fruit carry wild yeasts and other organisms on their skins. Fig 3.2 shows some of the surface organisms found on different foods. As long as the outer layers are intact, these organisms grow very little and do not pose a major health problem. Fresh meat, chickens and fish may carry the organisms which were present in, or on, the animal before it was killed, particularly gut organisms and those found on hair, hide or feathers. These are generally found on the surface of the food and don't invade deeper layers immediately. Table 3.1 lists some of the commoner organisms found on foods.

If food is exposed to the air – on market stalls, shop counters or kitchen worktops – it gradually gathers airborne organisms. Fungal spores are very common contaminants and houseflies have long been recognised as carriers of micro-organisms. Handling and processing add even more unwanted organisms; the more processing that occurs the more likely is contamination from people's clothes and skin. Micro-organisms can be transferred easily from one food to another, for example when fresh meat is stored next to cooked meats. This is called **cross-contamination**.

Some foods, such as wine and olives, are made using organisms found on their surfaces naturally. Conditions are manipulated to inhibit the growth of other organisms. Wines were traditionally made with wild yeasts found on the skins of the grapes. Different yeast strains are often confined to small areas and each gives a particular flavour and aroma to the wine which is characteristic of that region. However, most modern wineries have scientific culturing programs breeding a specific strain of yeast which gives consistency to the wine and to reduce souring. Fig 3.3 outlines modern wine production.

ORGANISMS ON FOOD

A tomato straight from the plant may have millions of organisms on each cm^2 of its surface. After washing, this reduces to a few hundred. Cabbage from the field has 1 to 2 million organisms per gram of outer leaf but after washing and trimming this falls to less than a quarter. Right in the centre of the cabbage there are only a few hundred organisms per gram.

QUESTIONS

3.1 What is meant by cross-contamination?

3.2 Explain why there are micro-organisms found in and on food.

3.3 List the ways in which food is made unacceptable by microbial growth.

3.4 Do you think there are many bacteria in each of the following foods?
(**a**) an orange
(**b**) a carton of pasteurised milk
(**c**) a vacuum-packed kipper
(**d**) roast beef, carved then left covered overnight
(**e**) a carton of natural yogurt.
When you have read *all* of this chapter look at your answers again. Alter your answers if necessary.

WINE PRODUCTION

The grapes used in wine making are all from the grapevine *Vitis vinifera*. There are a great many varieties of this plant each producing a different kind of grape and known by its varietal name, for example Pinot Noir, Merlot and Chardonnay. Each grape type has a slightly different chemical composition in its fruit which affects the wine it is made into.

The climate and the kind of soil also affect the growth, for example free-draining soils are warmer and the vines grow more quickly. Once the grape has ripened it will start to shrivel and lose water; this tends to increase the proportion of sugar in the grape. Attack by the *Botrytis* fungus has a similar effect making sweeter wines.

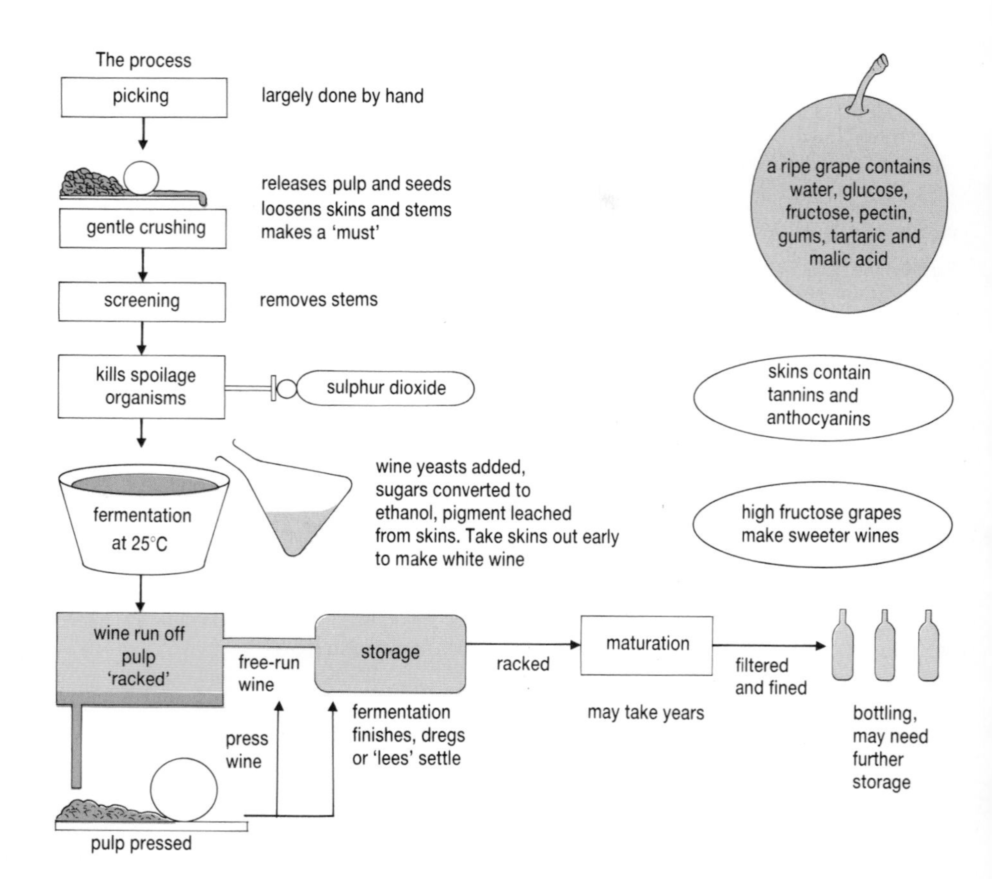

Fig 3.3 Wine production.

3.3 MAKING FOODS WITH MICRO-ORGANISMS

There are traditions of making foods using micro-organisms in many parts of the world. In the past, this was usually to preserve food for storage, to be used when supplies were scarce. Foods such as sauerkraut, dill pickles and the animal food silage were used in late winter and early spring when green vegetables were virtually unobtainable. Other foods were treated to bring out flavours as well as preserving, for example by fermenting soya beans. Foods which deteriorate very rapidly in warm conditions, for example milk, could be kept longer if made into cheese and yogurt.

The use of microorganisms to process foods involves the manipulation of growth conditions to select *for* the growth of the *desired* organisms and to *select* against the growth of others.

Though at the time people had very little idea of the biological processes involved, the traditional methods were based upon the manipulation of environmental conditions to bring about the preferential growth of certain organisms. Many modern techniques are merely large-scale refinements of the traditional methods, with the emphasis on the hygiene in factories, the maintenance of pure, stable cultures of microorganisms and the precise control of manufacturing conditions.

A wide range of foods are microbially processed in some way, including bread, beer, wines and spirits, olives, sauerkraut, yogurt, cheese, fermented milk drinks, buttermilk, vinegar, bean curds, sauces, coffee, tea and cocoa. Micro-organisms are also used to make large amounts of ingredients used in the manufacture of food and drinks. Citric acid used in soft drinks and sweets is made microbially, as are glucose syrups for baked goods and thickening agents for low calorie foods. Single-cell protein and mycoprotein is made from micro-organisms growing on wastes from other industries. They are used as animal feed or are textured for use in pies and other dishes; it is hoped that they will eventually help to fill the protein gap in the world food problem.

3.4 LACTIC ACID BACTERIA IN FOOD MANUFACTURE

Lactic acid bacteria are used in many food processes. They are fermenters that need high nutrient concentrations which limits them to nutrient-rich environments. They grow well in milk and foods rich in sugars, in vegetation, in animal intestinal tracts and on animal mucous membranes. They are used as natural contaminants or are added deliberately to the food during processing

Fig 3.4 Salami makers mix lactic acid bacteria with the meat ingredients before filling and tieing the salami skins. As the salami matures lactic acid inhibits the growth of spoilage organisms.

Olives

The preservation of green olives uses organisms living on the surface of the fruit in a carefully controlled environment. Olives are harvested when green, and bruised fruit are rejected as potential sources of spoilage. The fruit is soaked in alkaline solution for a time to remove the bitter substance oleuropein and then washed, after which they are put in barrels and covered in brine. This creates a low oxygen environment and only fermenters with high salt tolerance can grow, using nutrients leached from the olives. A succession of different species is set up in the first few weeks. First, *Leuconostoc* predominates and generates lactic acid, then *Lactobacillus plantarum* and *L. brevis* dominate. Few other organisms can grow in this environment. Eventually the pH drops to 3.8 with about 1 per cent lactic acid. Finally the olives are packed in brine, though if they are stuffed they are left for another month for the pepper to ferment.

Sauerkraut

Naturally occurring lactobacilli predominate in the production of sauerkraut too. To make sauerkraut cabbage is shredded, packed with salt and pressed. This leaches nutrients from the leaves which are rich in nitrogen and carbon compounds, but the environment created is virtually anaerobic. These conditions are ideal for lactic acid bacteria which are found naturally on cabbage leaves, though a starter culture may be used. *Leuconostoc* grows first but is superseded by *L. plantarum* over a period of weeks. Their activity quickly generates pH 5 or less which inhibits other organisms.

Cheese

Lactic acid bacteria live on the skin of cows and other dairy animals. They enter milk when the animals are milked, however most milk nowadays is pasteurised before being used for cheese making, and a starter culture of lactic acid bacteria therefore has to be added. Rennet, which is added to promote curdling, may be of animal origin, but is increasingly a microbial product.

Lactic acid bacteria require high levels of nutrients and produce lactic acid as a metabolite. Lactic acid inhibits the growth of other organisms and produces physical changes in the food.

Lactobacillus and *Streptococcus* species ferment lactose in milk to lactic acid. This lowers the pH and affects the structure of milk protein causing coagulation. The milk is separated into coagulated protein and fat curds

Fig 3.5 Fungal spores on the outside of a Stilton cheese germinate, but those in the centre cannot grow in the low oxygen environment. Piercing the cheese allows air to penetrate and encourages fungal growth, which matures and flavours the cheese.

and whey. The whey is drained off for animal feed. The manipulation of environmental conditions, together with the careful use of heat and salt, converts the curds into the various kinds of cheese. For example, heating to 50°C hardens curds and ensures that only certain heat-tolerant bacteria species thrive, whereas heating to 38°C will merely make a firmer curd ready for cheddaring.

Salt is then added, the curds pressed and the cheese is ripened. The ripening organisms are bacteria or fungi which carry out a range of proteolytic or lipolytic reactions. The by-products, which are organic acids (lactic, butyric, caproic, ethanoic and diacetyl acids) together with esters and amines, give characteristic flavours and aromas to the cheese. A method has been developed of growing the cheddar cheese-ripening organisms in vats and extracting the enzymes which mature cheese. These are added to the curds with salt before pressing and reduce the ripening time by about a half.

Cheeses with 'eyes', for example Gruyere, come from the action of *Propionobacterium* producing carbon dioxide which is trapped as bubbles, and propionic and succinic acid give the cheese its characteristic flavour. Blue cheeses acquire fungal spores from the atmosphere of the creamery, or the milk may be inoculated with spores at the beginning of cheese making. The fungal spores only germinate in the cheese if there is sufficient oxygen, so cheeses may be pierced with needles to allow air to enter the centre of the cheese, as illustrated in Fig 3.5. Rinds are formed by organisms wiped onto the outside of the cheese. Cheeses such as Brie and Camembert are made in this way. The organisms secrete enzymes which penetrate the cheese and produce flavour chemicals.

3.5 YEAST AND ITS PRODUCTS

The use of yeast in food production is probably the oldest form of biotechnology. Bread and fermented drinks have been with us throughout recorded history. Modern methods are simply refinements of the old techniques, as seen in Fig 3.6. There is more information of the biology of yeast cells in Chapter 1.

Fig 3.6 The techniques used 200 years ago of mixing, shaping, proving and baking are still used today; but modern bakeries use machines instead of hands to carry out these tasks.

Bread

When flour is made into dough, enzymes from the original grain act on starch to make a mixture of mono- and disaccharide sugars. Many bread flours now contain a fungal amylase to enhance the process. Yeast varieties of the species *Saccharomyces cerevisiae*, ferment the sugars as a respiratory substrate. Respiration generates carbon dioxide, making the dough rise as bubbles of gas are caught in the elastic protein pellicle. Water and ethanol are also made, but these are driven off by baking. *Lactobacilli* may grow at first, generating lactic acid. This contributes to the final flavour of the bread and inhibits other organisms. Bakeries may also add other ingredients such as whiteners, raising agents, stabilisers and flavourings.

Beer

Beer has a long manufacturing history and brewers jealously guard their particular brews. The varieties differ in the types and proportions of ingredients used and the treatment each ingredient undergoes. Brewing is quite a straightforward process, summarised in Fig 3.7. Beer is made from

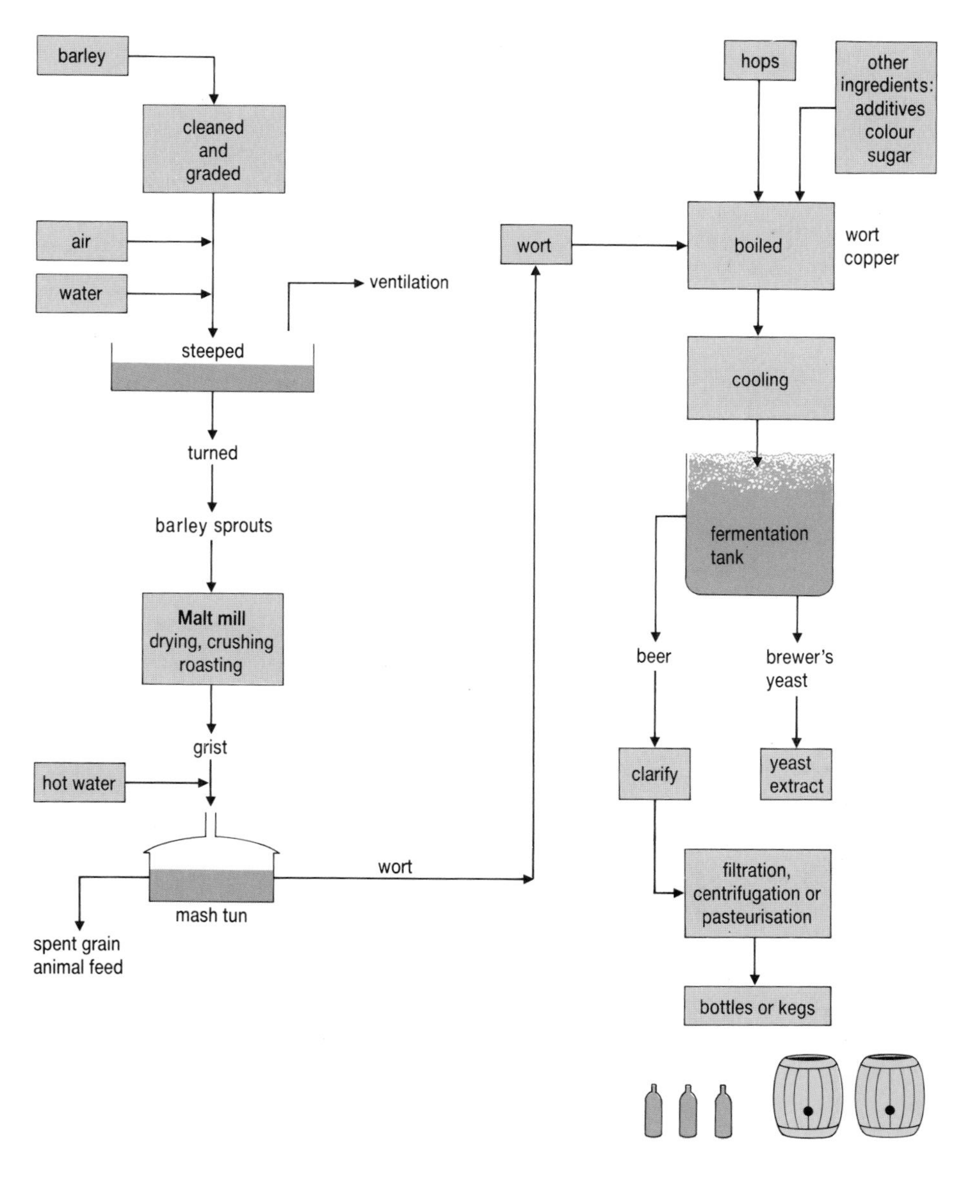

Fig 3.7 The brewing process.

Fig 3.8 Each batch of beer is fermented in a large deep tank; process workers use a walkway along the top of a row of tanks and need ladders to get inside them to clean between batches. The progress of the fermentation is monitored for ethanol production, clarity and quality.

barley grains, rice or corn which are moistened and allowed to sprout. During germination the starch reserves are converted enzymically to smaller molecules. Crushing and roasting the sprouted grains gives a darker colour and richer flavours. More enzyme action in the mash tun produces monosaccharides and simple nitrogenous compounds. The nutrient-rich liquor produced is called **wort**. The higher the sugar content of the mash the more alcohol is eventually made. Some brewers now add enzymes, such as amyloglucosidase and α-amylase, to the mash or fermentation vessel to get as much sugar as possible. This results in a low-carbohydrate or 'lite' beer. Hops, or sometimes just hop extract, and colourings are added and the mash is boiled. Hops give flavour and also have anti-microbial activity.

After cooling, the mixture is run into a deep fermentation tank and inoculated with brewer's yeast, strains of the yeast *S. cerevisiae*, usually saved from the previous batch of beer. The deep tank and the layer of carbon dioxide generated by the activity of the yeast makes the tank anaerobic and so fermentative metabolism is ensured. Ethanol is produced as a result of the anaerobic fermentation of sugars, in varying quantities but is seldom over 6 per cent. Once fermentation is complete the beer is separated from the yeast, clarified, treated to prevent spoilage and packaged.

Contaminants could establish quickly and hygienic conditions must therefore be carefully maintained if the beer is not to go sour. English beers are known as 'top brews' as the gas generated by the yeasts float them up to the top of the tank where most of the fermentation takes place. Many European beers are brewed with a variety, *S. carlsbergensis,* which is a bottom fermenter developed for long fermentations at low temperatures to make lagers. A simplified process is used to make spirits – the alcohol generated, together with flavour chemicals, is distilled off and matured.

3.6 ASPERGILLUS FERMENTATIONS OF SOYA BEANS

Soya beans are a staple part of the diets of people worldwide and a rich source of protein. They are versatile, in that they can be cooked or fermented in a variety of ways to make many foods; Table 3.2 lists some of them. In China soya beans have been used for over 3000 years, and China introduced 'chiang', a form of fermented sauce, to Japan over 1000 years ago. By the sixteenth century Japan had industrialised production of soy sauce, and was exporting by the seventeenth century.

Table 3.2 Traditional soy foods

Food	Source	Where used
miso paste	cereal and soya beans using *Aspergillus oryzae* and *Saccharomyces rouxii*	Japan and China; other Asian countries
natto	soya beans using *Bacillus subtilis*	Japan
Shoyu (soy sauce)	soya beans and wheat using *A. orzae, Lactobacillus, Hansenula* and *Saccharomyces*	China and Japan
Sufu	soya beans using *Actinomucor elegans* and *Mucor* spp	China and Taiwan
Tempeh	soya beans using *Rhizopus oligosporus*	Indonesia

(Adapted from Rackis, 1979).

Fig 3.9 Making Soy Sauce. The fermented mash is poured into filter cloths which are then shaped into parcels. The cloth parcels are stacked and pressed in a heavy press. The brown raw soy sauce runs down the side of the press and is collected for refining.

Soy sauce

Soy sauce is a food flavour and colour and has a high salt and peptide content. It is a very important ingredient in oriental cookery, particularly useful in bland dishes of fish, rice, bean curd (tofu), and boiled vegetables. In the UK we use much less but the chinese sauce market (70 per cent of this is soy) amounts to about £4 million per year, most imported from Hong Kong.

Traditional production

In the past, techniques varied little in different countries. In Korea soya beans were soaked to remove the anti-trypsin factor, cooked then pounded into a paste. The paste was made into small balls and left out to be inoculated naturally with air-borne spores of *Aspergillus* or *Rhizopus*, common saprophytic fungi. They were left for several months over the winter, and in the spring were extracted with salt water. The liquid produced was boiled and fermented in the sun to make soy sauce. The paste left over was mixed with salt and stored as miso, a soy food.

Industrial Production

The modern process uses the enzymic hydrolysis of beans and cereal proteins to amino acids and small peptides. The fungi *Aspergillus oryzae* and *Rhizopus* together with lactobacilli and yeasts provide the enzymes to break down proteins and starches. The type of sauce made varies with the ingredients and micro-organisms used. Soya beans are soaked and boiled and wheat is roasted and crushed. They are mixed together and inoculated with the fungus known as **Koji**, usually a culture of *A. oryzae*. The fungus grows for two or three days on the mixture, which has to be well aerated and spread out to prevent overheating. It is then mixed with saltwater to make a mash. The salt content is regulated and the mash left in deep, cool fermentation tanks for several months (see Fig 3.9).

The conditions in the fermentation tanks encourage the growth of lactobacilli and yeasts. Vigorous lactic acid and alcohol fermentations take place and the mixture is aged to make raw soy sauce. This is filtered and clarified and the sediment is pressed to use as animal feed. The filtrate is pasteurised, which gives more colour and a stronger flavour, and the sediments and oils removed. This refined sauce, shoyu, can then be bottled.

3.7 VINEGAR MAKING

Traditional vinegar production used to be a slow process; wine was put into barrels and acetic acid bacteria developed as a skin on the surface of the wine. After several weeks the wine was converted to vinegar. Modern vinegar making is one of the few microbial processes which is run as a continuous culture system. It uses a mixture of *Acetobacter* species which require aerobic conditions for growth and often grow as thick clumps held together by capsular material. They partially oxidise ethanol to ethanoic acid (acetic acid) in respiration which makes the sharp, sour taste of vinegar. The other flavours come from other molecules such as esters and volatile oils made by the micro-organisms and from the starting material.

The starting material can be wine, beer or a liquor from the fermentation of fruit juices by *Saccharomyces cerevisiae* specially for the process. The oxidation needs high oxygen concentrations so the wine or beer is trickled over a finely divided material, traditionally wood shavings, coated with a film of bacteria. Fig 3.10 illustrates the process. Aerobic conditions are ensured by blowing air from the bottom of the tank through the shavings. Within four or five days the ethanol is converted to ethanoic acid. The process can be speeded up if a portion of the raw vinegar drawn off from the bottom of the tank is passed through the process again to reduce the ethanol content and increase ethanoic acid; the rest of the raw vinegar is matured. The system is cooled by cooling coils inside the tank or by the air blown through.

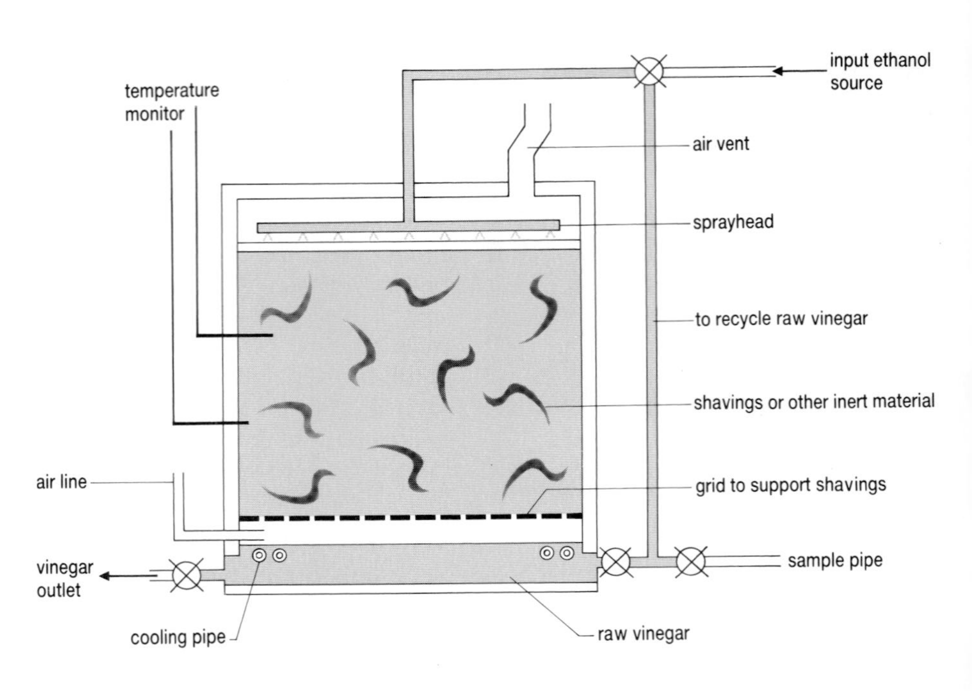

Fig 3.10 Vinegar manufacture.

This method for making vinegar is being replaced by technology developed for the growth of fungi. The bacteria are grown as a submerged culture in an ethanol source which is stirred vigorously and aerated. This system results in a very high rate of conversion of ethanol.

3.8 SINGLE-CELL PROTEIN

The world shortage of protein foods has been a problem for many years. The main protein supplement is soya meal which most countries import and use to supplement animal foods. However, using grain and protein to feed animals to make protein is wasteful; it makes more sense to use protein for people directly rather than passing it through inefficient animal metabolisms first.

Algae and *Spirulina* are gathered for food in some parts of the world but the habit isn't widespread. The idea of using micro-organisms to make protein-rich foods has been around for some time but it has taken a lot of time and money to develop the techniques needed. Cheap materials are needed as a substrate and the organism has to be safe to eat. The product must be easily processed into an acceptable food which must not have detrimental effects on the consumer. There must be no residues or contaminants from the substrate nor must any uncommon constituents cause long-term damage.

The first attempts at protein foods caused problems as they had a high nucleic acid level which led to kidney problems; processing now eliminates excessive amounts of nucleic acids. The processes which have been developed use a variety of micro-organisms growing on different substrates. The cells are harvested and processed to make **single-cell protein**, or **SCP**. SCP was used first for animal feed and is now a major product. The biggest problem now is how to persuade people to eat foods made from micro-organisms so that the manufacturing companies can recoup their research costs; though they may have been happily eating mushrooms, yogurt and cheese all their lives. In Britain any company which wishes to introduce a new food, which is made of substances not usually used for human consumption or by new processing techniques, has to clear

Fig 3.11 This pie is made of mycoprotein.

it with the Advisory Committee on Irradiated and Novel Foods and the Ministry of Agriculture, Fisheries and Food. Many other countries operate similar schemes.

Fungal SCP

Many of the processes for manufacturing SCP use wastes which could not otherwise be used as food and convert them into food supplements for animals. If a substrate is available and the fermentation can be done cheaply enough, small process plants are viable. These processes can also solve local pollution problems. An early process developed in Finland uses carbohydrate-rich wastes from wood and paper industries, which used to be a source of water pollution, to grow a fungus. *Paecilomyces varioti* is grown in continuous culture to make protein, sold as Pekilo™. In Britain a similar process uses flour waste from flour making to grow a fungus *Fusarium*. The mycelium is harvested and processed to make fibres which are pressed together to make a material called mycoprotein. The mycoprotein has a high protein and fibre content but no cholesterol. The texture is made to resemble meat chunks and it can be flavoured to taste like beef or chicken. It is used in pies and other products and is sold under the name Quorn™. Fig 3.11 shows a pie made from Quorn™.

QUESTIONS

3.5 Make a summary of the use of a named micro-organism in food production.

3.6 Silage is made when the leaves of plants such as sugar beet, grass, maize and lucerne are compacted in silos with molasses and some mineral acid.
(a) What is the likely purpose of (i) the molasses, and (ii) mineral acid?
(b) Which sorts of bacteria are likely to grow in the silage?

3.7 Flow diagrams have been given in this chapter to illustrate wine and beer making. Construct a similar diagram for cheese making.

3.8 Construct a chart of the food processes described in detail in this chapter and for each say whether the process is a continuous culture or a batch process and if it needs aerobic or anaerobic conditions.

3.9 **(a)** Briefly survey the use of micro-organisms in the processing of food products.
(b) Describe in detail **one** process in which a micro-organism plays a substantial part in the formation of the final food product.
(JMB)

3.9 FOOD SPOILAGE

It is the capacity of micro-organisms to exploit ecological niches which causes spoilage problems. There are micro-organisms which can flourish in the most adverse environments and may thrive with the lack of competition in a jar of jam or vacuum packed bacon. Very few of the organisms found on the surface of food are capable of rapid growth, but those which can will grow, and their activity may change conditions so that others can grow too, including spoilage organisms.

Spoilage organisms may not have been dominant in the original food flora but are capable of rapid growth and exploitation of the food source.

Many foods are resistant to spoilage because they have protective outer layers, or contain anti-microbial chemicals such as the lecithins or lysozyme. If microorganisms get in and conditions are favourable for their growth then spoilage is rapid. Damage from other causes may allow microorganisms to enter foods. Many foods suffer **autolysis**, that is breakdown due to their own enzymes. Physical damage from chopping and processing foods encourages autolysis which is undesirable, for example prepared salads will become slimy and limp. Even sterile food autolyses unless it is treated to denature enzymes. The damage done makes it much easier for micro-organisms to enter which would not have been able to penetrate a 'whole' food. Different organisms can grow on different foods and cause different effects, as shown in Table 3.3.

Table 3.3 Some specific spoilage organisms

Food	Damage	Organisms
wine, beer	souring due to ethanol oxidised to acetic acid	*Acetobacter*
peanuts, cereals, fruit, dried foods, bread	aflatoxin produced causing food poisoning, mycelial growth through food	*Aspergillus* spp
pasteurised milk	'bitty milk' from lecithin breakdown	*Bacillus cereus*
most foods	green rot at low temperature	*Cladosporium*
pickles	blackening as SO_4^{2-} reduced	*Desulphovbrio*
onions, potatoes	soft rot due to pectinase production	*Erwinia*
dairy products	felty red, yellow, orange colonies	*Geotrichum*
citrus fruits, cheese	soft rot, green growth	*Penicillium*
bread, fruit, vegetables	visible mycelial growth	*Rhizopus*

In animal-derived food, the problem is not so much unpalatability but the presence of pathogens. These are most likely to have come from the gut of the animal. Meat, as a high protein food, is spoiled by a number of organisms; *Pseudomonas, Micrococcus, Bacilli,* and *Proteus* particularly. Some are psychrophiles which make meat go green or black in cold storage; not necessarily harmful but aesthetically unacceptable. Shellfish carry micro-organisms from the waters they grow in, particularly enteric organisms from sewage contaminated water. Great care has to be taken maintaining water quality and in cleaning shellfish before they can be eaten.

Some organisms produce **enterotoxins**, for example *Staphylococcus pyogenes aureus,* but most food poisoning incidents involve *Salmonella* species. The pattern of food poisoning infections is illustrated in Fig 3.12. Extra care is taken with poultry as these are common sources of *Salmonella.* Chickens are chilled very quickly after death which reduces the growth of micro-organisms. *Salmonella typhimurium* is by far the most common species involved, with cold roast beef and chickens, reheated stews, steak pies and incompletely cooked chickens and eggs being the most frequent sources of illness. Details of salmonella infections can be found in Chapter 8.

Potential pathogens are unlikely to be found on vegetables unless they have been irrigated with untreated sewage or handled in an unhygienic way. Organisms such as *Erwinia* and *Lactobacilli* cause rots in stored vegetables and fruits. There is, however, a health hazard in stored plant material. The fungus *Aspergillus flavus* and related species can grow on moist, stored grains and nuts, producing a toxin, **aflatoxin**, which has been found to cause deaths in humans and other animals.

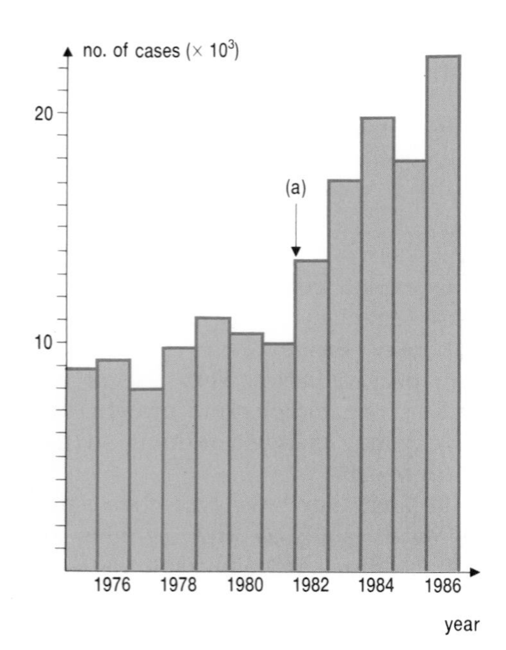

(a) A new category of cases added 'cases ascertained by other means'.

Fig 3.12 Food poisoning cases in England and Wales.

AFLATOXIN

In 1959 thousands of young turkeys died on turkey farms in Britain. There were also losses of ducklings, pheasant chicks and young partridges. In other parts of the world there were similar losses: trout in California and ducklings in Africa. All seemed to be due to a toxin. The young turkeys had been fed on pellets of peanut meal and other materials. When the meal was examined microbiologically it was found to contain many organisms including several fungi. One was identified as *Aspergillus flavus* which produces the toxin aflatoxin. Examination of aflatoxin has shown it to be a mixture of compounds, each of which is toxic.

The toxin affects a wide range of living things, including other micro-organisms, plants and animals. The dose needed to cause damage is very low and young animals are far more susceptible than older ones. The toxin has a variety of effects; it makes capillary walls fragile leading to haemorrhage in the tissues, hepatitis and death. It is also a potent carcinogen affecting the liver, kidneys and colon. The effects in humans are less well understood as the only information is from people with accidental poisoning, but in one outbreak, caused by mouldy maize, 106 people died and 291 had liver damage.

The fungus needs to grow in high humidities and high carbohydrate concentrations to make the toxin as a secondary metabolite. A wide variety of foods provide the right conditions including cottonseed, corn, peanuts and other nuts and their products. The fungus grows best if the crop has been damaged during harvesting or if there has been insect damage in storage. Other fungi growing in the same products compete with it, reducing its growth and detoxifying the toxin. When animals are fed on meal carrying the toxin, it accumulates and is passed along the food chain, turning up in eggs and milk.

The fungus is controlled by better harvesting technology and storage techniques. A carbon dioxide-enriched atmosphere reduces growth of the fungus. Sorting contaminated grain isn't very easy nor is it satisfactory as the toxin can diffuse into apparently uncontaminated grain. Products such as peanut oil can be decontaminated but the techniques cannot be applied to all products. Foods contaminated with aflatoxin have to be destroyed. Detecting aflatoxin is difficult. Currently, samples of food are subject to gas chromatography, but a diagnostic kit using monoclonal antibodies to aflatoxin is being developed.

QUESTIONS

3.10 Study the section on aflatoxin and answer the following questions:
- **(a)** Why do we have evidence of human effects only from accidental poisoning?
- **(b)** What do the terms (i) secondary metabolite and (ii) carcinogen mean?
- **(c)** Why are there more animal deaths than human deaths?
- **(d)** Suggest a harvesting and storage regime which could be adopted by a peanut farmer in the southern part of the USA, where the weather can be hot and humid, which would minimise loss due to the fungus *Aspergillus flavus*.

3.11 Using the information in Table 3.3 and the text, make a table of the main kinds of protein foods and their spoilage with the causative agent.

3.12 Use Fig 3.12 to describe the general trend of food poisoning cases. The arrow indicates a change in the data recorded; after this date numbers of cases 'ascertained by other means' were included with formal notifications by doctors. Do you think that the number of cases from 1975 to 1981 is an accurate record?

3.10 PRESERVING FOODS

The aims of preservation are to prevent food spoilage by autolysis, to kill spoilage organisms or prevent their growth, and to prevent further contamination. Whichever method of food preservation is used there will be an effect on the food. For example heat treatment alters the flavour, taste and texture of foods. It also adds to the cost of preparing food for sale. Different methods work in different ways. Some methods kill bacteria – they are **bacteriocidal**; others are **bacteriostatic** and merely inhibit growth. Some methods combine both bacteriocidal and bacteriostatic practices.

Canning

Preserving food in cans was started over 150 years ago and it revolutionised food storage. The food is packed in cans, heated to a high temperature, sealed and then subject to sterilising temperatures. After the can is sealed no organisms can enter. The high temperatures needed are produced by using steam under pressure. Table 3.4 shows their effectiveness. Moist heat is far more effective at reducing the bacterial count than dry heat. Most organisms will die if exposed to this treatment, and even the most resistant spores will succumb after 20 minutes at 121°C at pH 7. Time in the steriliser can be reduced if the food is acidic, which inhibits the growth of most spores. For example, tomatoes at pH 4 need only a few minutes at 100°C.

Table 3.4 Time taken to reduce bacterial populations by 90%

Organism	105%	120%	130%	140%	150%
Bacillus cereus	12.1	4.2	2.6	1.3	1.0
Bacillus subtilis	27.8	4.5	3.1	2.1	1.1
Bacillus stearophilus	2857.0	38.6	8.8	3.9	2.4

After Miller & Koudler (1967) .

Cans of food have to be cooled in water. At this stage extreme care must be taken that there has not been a failure in the seam weld of the can, otherwise the cooling water can be drawn into the can as the contents cool and contract, which may carry organisms capable of growing in the can. Cooling waters are usually chlorinated and manufacturers inspect samples of cans for failures or dents which could also cause seam failures.

The storage life of food in cans varies according to content. Meat lasts for years, though oxidation of fats may eventually cause some changes in flavours. Fruits have a shorter storage time as the acid content can corrode the can. Not all foods can be canned because of the effect of heat on the food may make it unacceptable.

Freezing and refrigeration

Most micro-organisms grow very slowly at low temperatures. Lowering the temperature of food below freezing point stops the multiplication of any organisms in it and reduces their viability. Most foods can be frozen successfully, providing it is done carefully. The texture of food is spoilt if large ice crystals form, but if food is quick-frozen, that is it falls from 0°C to –4°C in less than 30 minutes, then small ice crystals are formed which do less damage. There are three ways of freezing food commercially. It can be passed between two very cold plates, through a blast of cold air, or dipped in liquid nitrogen. The last method isn't a practical proposition in most cases as it is very expensive. In freeze-drying, food is put in a vacuum chamber after freezing so that water sublimes off.

At the time of freezing the food will have all its original complement of microorganisms plus any it has acquired during handling. A few species can continue to grow, though slowly, at low temperatures. These psychrophiles include the spoilage bacteria *Micrococcus* and fungi *Cladosporium, Penicillium* and *Monilinia* which may grow while food is in storage. When food is thawed organisms will multiply rapidly. There will also be deterioration through autolysis which will be more rapid than usual because of ice crystal damage. Accidental defrosting, for example as a result of power failure or food being stored above the load line in a commercial freezer, will result in the rapid growth of micro-organisms present while the temperature is warmer. Very high bacterial counts will be reached if the food is refrozen and then defrosted again. A domestic refrigerator holds food at 4°C or lower. At this temperature microbial growth is reduced, prolonging the storage life of perishable foods such as salads and milk for a few days.

Fig 3.13 Farmers may pasteurise their own milk but much is pasteurised after collection. The milk passes in a thin stream between metal plates where it is heated then rapidly cooled before bulk storage.

Pasteurisation

Pasteurisation uses moist heat but the temperatures are much lower than in canning. Vegetative cells of many fungi and bacteria are killed if they are held at temperatures around 60°C for any length of time. In general, the higher the temperature the shorter the time needed to kill vegetative cells, however spores and some highly resilient species are not necessarily killed by this treatment. The method is most useful for foods which would be spoilt by the high temperatures needed for sterilisation.

Milk is pasteurised at 72°C for 15 seconds then cooled rapidly. This kills many spoilage organisms, for example the non-sporing *Lactobacilli,* also some potentially dangerous pathogens transmitted through milk (see Fig 3.13). For example, *Mycobacterium tuberculosis* which causes TB, was frequently carried in cows' milk in the past. Pasteurisation merely delays milk spoilage; curdling still occurs because of the activity of other organisms such as *Streptococci* and spore-forming *Bacilli.* If pasteurisation is combined with refrigeration, milk still keeps for several days. **UHT** milk is not pasteurised; it is sterilised. It has superheated steam blown through it at temperatures of 135–160°C for one or two seconds. This kills both vegetative cells and spores.

Other dairy foods such as liquid ice-cream are pasteurised and so are fruit juices and beer. The latter may already be acidic enough to inhibit most spores but it may still be pasteurised as well.

High osmolarity

Many foods are preserved using salt or sugar. This generates a high osmolarity in the food which most organisms find difficult to grow in. Jams and conserves are 50–70 per cent sugar and are boiled before packaging into hot jars. This sterilises the jam which is then sealed to prevent further contamination. Once the jam is opened, bacteria still cannot grow but some yeasts and fungi are able to tolerate the high sugar content. Both salt and sugar absorb moisture from the atmosphere and this accelerates spoilage as the concentration of sugar or salt falls.

Salting and brining uses salt concentrations of 20-30 per cent and is used for meat, fish and vegetables which are not cooked before brining. There are some halophytic organisms which can tolerate these high concentrations and cause spoilage.

Drying

Dried food has a very low free water content which reduces microbial growth. Dried cereals, grains and fruit, which have been dried by blowing hot air over them, can be kept for a long time. If the atmosphere becomes moist, the food absorbs moisture and the free water content rises, allowing microorganisms to grow.

(a)

(b)

Fig 3.14 **(a)** Scottish Arbroath Smokies are fish which are split and smoked overy a fire in a smokery to preserve and flavour them.
(b) In Chad the sun is hot enough to dry fish.

Smoking

Smoking is another traditional method of preserving fish and meat, shown in Fig 3.14. Food is suspended over a fire and hot smoke permeates through it. This kills many micro-organisms, denatures autolytic enzymes, and dries out the food. Many of the components of wood smoke are bacteriocidal or bacteriostatic, including cresols, organic acids and aldehydes such as formaldehyde. In some cases the food is salt cured first. If smoked foods become moist they may develop fungal growths and fats may become rancid.

Preservatives

There are a number of substances which can be added to food to stop the growth of micro-organisms and autolysis, and to prevent chemical oxidations. Collectively these are known as preservatives. Some of the processes already mentioned could be put in this category, for example smoking or salting food. Vinegar, often as pure ethanoic acid, is added to pickles, mayonnaise, sauces and coleslaw as a preservative as well as a flavouring because the acidity inhibits the growth of most organisms. Other acids such as citric acid and sorbic acid can be used in the same way. Within the EC, food additives are known by their EC code number and are printed on the labels of foods containing them. Table 3.5 shows some of the more common anti-microbial additives. Some foods, such as fruit, may be coated with biphenyl or o-phenyl phenolate to delay spoilage.

Table 3.5 Some common preservatives

EEC code number	Preservative	Uses
E200	sorbic acid	dried fruits, low fat spread
E210	benzoic acid }	foods made with fruit pulp
E211	benzoic acid salts }	soft drinks
E220	sulphur dioxide }	dried fruit, wines, pickles
E221–7	sulphur dioxide compounds }	preserves vitamin C
E250–2	sodium nitrates and nitrites	meat
E260	ethanoic (acetic) acid	pickles, sauces, mayonnaise
E270	lactic acid	soft cheese products
E280	proprionic acid	
E296	malic acid	mincemeat
E297	fumaric acid	

Radiation

Though radiation has long been used to sterilise and preserve such things as culture media, medical supplies and drugs it is not allowed for food preservation in many countries. It is used particularly to inhibit the growth of micro-organisms in chickens and shellfish and to stop sprouting in grains and vegetables.

Case study: The use of gamma radiation in food preservation

Gamma radiation has been used to sterilise materials for some time. Irradiation is usually from the isotope ^{60}cobalt, which is cheap and relatively easy to handle, and from ^{137}caesium. The dose required to kill organisms varies: 10 kGy for bacterial and fungal cells; spores take 10–50 kGy. Viruses are very resistant, needing 30–50 kGy. In contrast, organisms as complex as humans are killed with doses of only 0.06–0.1 kGy which causes damage to bone marrow cells. Generally a dose of 45 kGy is regarded as a sterilising dose. Radiation can be used on foods as a means of sterilising them, or as a partial steriliser in which spoilage organisms are destroyed but others may still be present.

Gamma radiation has other effects as well as killing spoilage organisms. At 20–150 Gy the sprouting of onions and potatoes after harvest is inhibited, extending their storage time. The market life of fruit can be extended by upto 20 days as ripening is delayed. In Canada, levels of 750 Gy have been approved for grain storage; normally upto 25 per cent of stored grain is lost by insect damage. This dose kills adult insects, larvae, eggs and pupae. The radiation leaves no residues nor is there reinfestation from insect eggs. Meat and fish can harbour parasites which are killed by radiation too.

Irradiation could be useful for controlling food-borne illness because food can be treated after packaging and sealing, preventing any further contamination. There is some risk, however, with partial sterilisation as some spores remain and these may cause spoilage which a consumer may not recognise.

The loss of nutrients in radiation-treated food is similar to the loss in foods which have undergone more conventional forms of heat treatment. A big worry for many people is the induction of radioactivity in food when it is irradiated. Radioactive sources with energy levels below 10 MeV must be used to ensure there is no induced radioactivity. When ^{60}cobalt is used, induced radiation can't be detected. Neither does the radiation affect chemicals present in the food, nor is there conversion to toxic compounds if doses below 50 kGy are used. Studies have been done to test whether radiation-treated foods are carcinogenic. As yet the results are inconclusive.

The main disadvantage of using irradiation as a method of preserving food is that it is very expensive. Its main advantage is that it can reduce losses of food and preserve high quality food, but it will not improve poor quality food. In one way it is unique, as it is the only method available at the moment for eliminating *Salmonella* from frozen meat and bulk animal feeds. This use of irradiation is permitted in some countries, for example The Netherlands, but not in most others, including the UK.

QUESTIONS

3.13 What is the difference between the terms 'bacteriocidal' and 'bacteriostatic' ?

3.14 Review the preservation methods given in this chapter and list them under three headings: 'bacteriocidal', 'bacteriostatic' and 'both'.

3.15 Why is it considered unwise to buy (i) dented tins and (ii) food for freezing which has already been frozen and defrosted once?

3.16 Many hospitals are adopting a 'cook-chill' technique of preparing meals in advance followed by reheating on the ward. Why is this technique thought to be better than cooking then holding in warming trolleys?

3.17 How do fungi spoil food? How may such spoilage be reduced?

3.18 Select one bacterium and one fungus which are of importance in food spoilage. For each,

(a) explain how the organism can be a serious cause of food spoilage,

(b) account for the ways in which the organism enters, survives and flourishes in food stuffs,

(c) describe how infection of the food can be prevented or suppressed. (*JMB*)

3.19 Review this chapter and examine the case study on the use of radiation for food preservation.

(a) List the points for, and the points against, the use of radiation on food.

(b) You are a member of the Meat Pie Manufacturers' Association (MPMA) in favour of radiation and are to present evidence to a government committee considering legislation on the use of radiation on foods. State your case.

(c) You are now a member of Clipshire Consumers' Group (CCG). What safeguards would you expect the Government to include to protect the general public?

You may prefer to do this question with a group of fellow students. Elect one person to be chairperson, and let two people prepare the case for the MPMA and two for the CCG, everyone else is a Member of Parliament on the committee.
What decision did you come to?

SUMMARY

The processes which micro-organisms use to obtain nutrients from food sources bring about chemical and physical changes in those foods. These changes may be desirable, for example in cheese production, or undesirable, as in the spoilage of food. Micro-organisms in foods are contaminants from the production areas or from processing, or they are added deliberately. Foods such as sauerkraut are made by manipulating the environment to encourage the growth of the desired organisms and inhibit the others.

Micro-organisms can grow in foods and cause spoilage, or generate toxins. Bacteriocidal preservation methods kill all the organisms present, whereas bacteriostatic methods reduce their growth rate. There is controversy over the use of radiation as a method of preservation.

Chapter 4

MICRO-ORGANISMS AND WASTE DISPOSAL

LEARNING OBJECTIVES

After studying this chapter you should be able to:

1. understand the problems caused by inadequate waste disposal;
2. describe ways in which micro-organisms are used to provide clean water;
3. explain how micro-organisms are used to convert waste materials into useful products;
4. describe how micro-organisms can be used to reduce environmental pollution;
5. describe some of these processes in detail.

4.1 THE PROBLEMS OF WASTE DISPOSAL

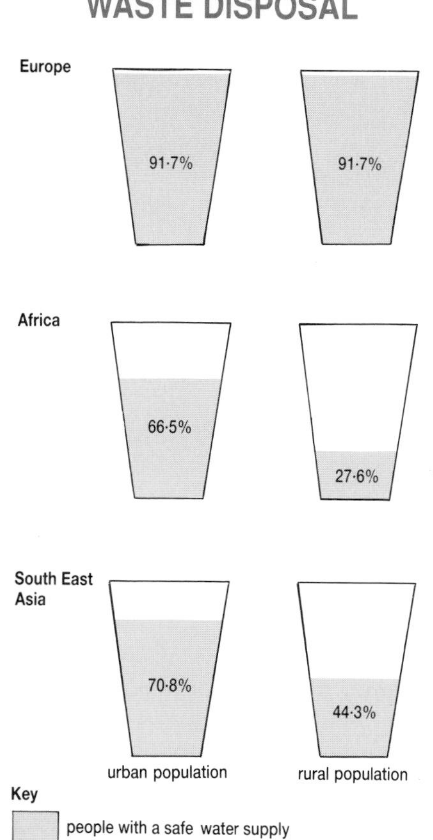

Fig 4.1 The world's safe water supplies.

A major change in people's lives over the last hundred years is the increase in the standard of health in developed countries. Improvements in individual and community hygiene have resulted in fewer people dying of disease and more people living longer. In the developed countries this virtual freedom from disease is taken for granted, but in many other parts of the world this is not the case. In these areas, the most vital ingredient in improved health is missing – clean water, uncontaminated by harmful organisms, resulting in high death rates from water-linked diseases such as typhoid, diarrhoea, cholera, guinea worm and river blindness. In 1980 it was estimated that about 50 000 people, many of them young children, died of water linked diseases each day. Many more were ill with these diseases, placing a great strain on resources. The World Health Organisation estimated that in 1985 worldwide 40 per cent of city dwellers and over 80 per cent of people in the countryside had no proper sanitation; only 38 per cent of the whole population of Africa had access to safe water. The 1980s were designated the decade of clean water by the United Nations, but sadly much remains to be done: it is estimated that nearly two thousand million people will still be without adequate sanitation by 1990. Fig 4.1 illustrates the situation in 1986.

Many waste materials enter the environment. Most are degraded slowly by living organisms into smaller harmless molecules. However some are not easily broken down, and instead they accumulate to levels which could pose health hazards or be offensive. There are new compounds, synthesised in the last 50 years, which are difficult to dispose of safely. These, and more traditional wastes, are dumped in the sea or incinerated to dispose of them but we now know that this is not a satisfactory long-term solution. For example, low levels of mercury, lead and other metals and some pesticides are taken up by living things from the environment. These are transferred to higher organisms along the food chain, in which they accumulate in harmful, and sometimes fatal, concentrations. Research into safer disposal of harmful chemicals and pollution prevention is being carried out as traditional methods of disposal are found to be too expensive or inadequate for modern safety standards.

4.2 CLEAN WATER

Natural waters

As rain falls it absorbs carbon dioxide and other gases and collects dust from the air. More substances are collected as it flows over the ground. Salts dissolve in water as it percolates through soil. By the time water enters rivers, it is carrying mud, vegetation fragments, grit and soil micro-organisms.

The number of micro-organisms in natural waters depends on several factors. The physical environment and the nature of the other inhabitants of the ecosystem are important but the most significant factor is nutrient availability. Open water usually has few nutrients, restricting the numbers of micro-organisms that can survive there. Most nutrients are found around solid objects such as dead wood and stones on the bed of the pond or lake where sediments and particles of organic material lodge in crevices. Other inhabitants of the aquatic ecosystem such as water insects, crustacea and molluscs also live among the stones and weeds. Many micro-organisms feed saprophytically by secreting enzymes into their immediate surroundings to break down organic material; the soluble molecules produced are then absorbed. In open water these enzymes are quickly dispersed and diluted, but in the crevices they work more effectively.

Micro-organisms are found in aquatic ecosystems but there are few except in nutrient rich areas.

The micro-organisms present are often soil saprophytes such as *Streptomyces* and *Bacillus* species that get washed into watercourses by rainwater, though there may be specialised organisms in acidic or alkaline waters. Very deep or very turbid waters may be anaerobic near the bottom; the only micro-organisms living there are those aquatic species which can respire anaerobically.

(a)

(b)

Fig 4.2 Rivers with little pollution **(a)** have a good growth of vegetation on the banks, the water is clear and houses a complex ecosystem of animals, plants and micro-organisms. Water which is subject to pollution **(b)** has a little vegetation in the water or on the water margins, may be discoloured or turbid and has little or no life in it.

Polluted waters

When polluting materials enter the water, changes take place in both the physical nature of the water and the balance of living organisms found there. If the water is polluted with organic matter then micro-organisms which are not normally found in water may flourish and populations of all bacteria may rise from 10^2 per cm^3 to 10^6 per cm^3. Contamination in natural waters can extend for many kilometres downstream of the entry point. Until recently the water at the mouth of the River Rhine carried over 10^5 organisms per cm^3 as a result of the accumulation of thousands of tonnes of salts and thousands of litres of oil wastes from entry points in three different countries.

Farms are sources of two important types of contamination: nitrate pollution and waste from farm practices. Nitrate fertilisers are used extensively to grow crops but almost half of the nitrates added to the soil are leached into water courses, unused. The excess nitrate disturbs the ecological balance of water organisms. Animal wastes such as slurry, which is manure diluted with water, can cause pollution if it is allowed to enter water courses. Dairy farmers store grass in silage clamps to feed cows in the winter. Water draining through these clamps gains acids from the fermentation, and unless this is adequately treated it is one of the most polluting wastes, being four times as polluting as piggery waste (see Fig 4.3) .

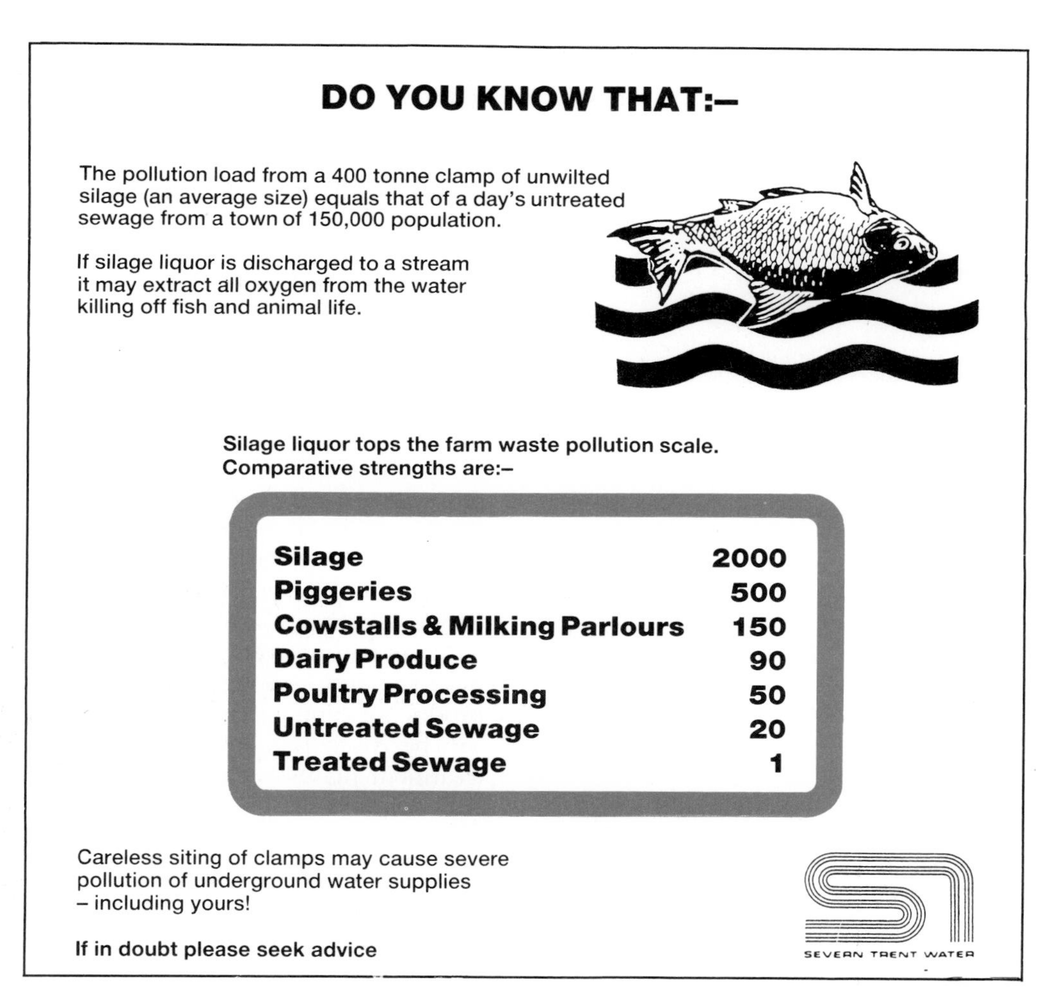

Fig 4.3 The relative strengths of farm pollutants.

There are four important ways in which pollutants can affect watercourses.

Organic waste

Sewage and other organic wastes cause a fall in levels of dissolved oxygen as the waste is oxidised to carbon dioxide, nitrate, phosphate and other compounds. There is increased algal growth in surface waters brought about by the increased amount of nitrate, leading to eutrophication (section 1.12). When the oxygen content of water falls, aerobic organisms such as fish suffer and anaerobic organisms flourish. Anaerobe activity generates acids, sulphides and amines which affect other aquatic organisms. Organic chemicals can also cause unpleasant smells or have toxic effects on the water inhabitants.

Particles

Small particles and silt suspended in the water from mining, quarrying and other industries block light transmission through the water, reducing photosynthesis in aquatic plants. The particles also clog the gills and delicate body structures of aquatic animals. Heavy sediment deposits hinder river inhabitants trying to attach themselves to the substrates and they are washed away.

Harmful micro-organisms

There may be disease-causing organisms in water carried by untreated sewage; examples are cholera and typhoid bacteria, polio virus and the eggs of parasitic worms. If contaminated water is used for drinking, washing or preparing food these harmful organisms can be passed on to humans and cause disease.

Heat

Many industries use water as a coolant. Cooling water is warmed in the process and discharged into rivers where hot water adversely affects the inhabitants. Micro-organisms can grow in a wide range of temperatures but most cannot survive the sudden changes in temperature which can occur near cooling water outlets.

Contaminated waters are aesthetically unacceptable, carry disease-causing organisms and disturb aquatic ecosystems.

Water treatments are designed to reduce the amount of contamination entering natural waters, to reduce the number of harmful organisms and to produce an aesthetically acceptable watercourse. In Britain, industries producing contaminated waste water, or **effluent**, have to treat it before it can be discharged into natural waters. Most used water goes through treatment plants such as sewage works which remove offensive material and degrade organic compounds to simpler chemicals such as nitrates, phosphates and carbon dioxide, a process called **mineralisation**. Water used for drinking water is monitored and treated before it enters the distribution network.

QUESTIONS

4.1 Suggest four sources of water contaminants.

4.2 The numbers of micro-organisms in natural waters are usually low. Why do the numbers increase if untreated sewage is discharged into water?

4.3 Explain the meaning of the term 'mineralisation'.

4.4 Read the information about the Public Health Authorities on page 74 then answer the following questions.
- **(a)** What to you think is meant by 'the economic cost of sickness'?
- **(b)** What is a miasma?
- **(c)** Who will pay for the cost of emptying privies and ashpits of Dudley householders? (Fig 4.4(b))
- **(d)** Why do you think the regulations that each house should have a water closet, a cesspit or an ash pit led to improvements in the health of urban inhabitants?

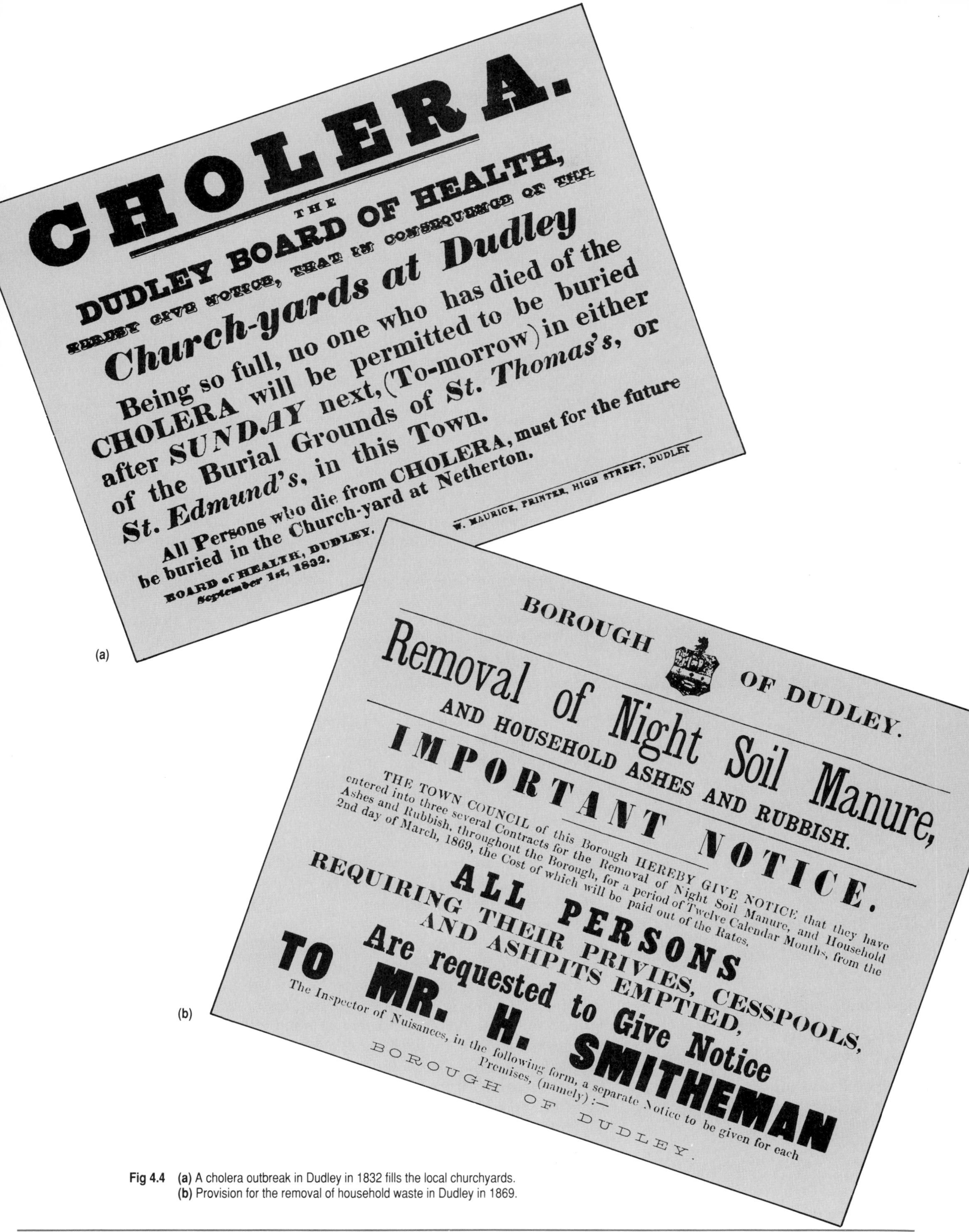

Fig 4.4 **(a)** A cholera outbreak in Dudley in 1832 fills the local churchyards.
(b) Provision for the removal of household waste in Dudley in 1869.

INTRODUCTION OF PUBLIC HEALTH AUTHORITIES IN BRITAIN

At the beginning of the nineteenth century there was no regular system to dispose of rubbish in Britain. The usual way of getting rid of water and sewage was to pour it into the local brook, which also supplied drinking water. Life expectancy was about 40 years and sickness and disease were rife. In 1831 an outbreak of cholera, referred to in Fig 4.4, caused a Central Board of Health to be set up which recommended providing a system of public health authorities. These authorities were set up but were rather ineffectual. In 1838 there were major outbreaks of disease and investigators concluded that the physical environment of the poor was largely to blame. A report on conditions was published stressing the economic cost of sickness in the 'labouring classes' and the effects of bad housing and sanitation on health.

Little was done at first but by 1848 the first Public Health Act was passed and Medical Officers of Health were appointed. Houses had to have a water closet, a cesspit or an ash pit. Weekly sickness returns were compiled and information was systematically reported. At this time people were unaware of the true nature of infectious disease; they thought that it was caused by miasmas in the air. Consequently the main aim was to provide clean healthy surroundings where these miasmas did not develop. Fortunately the sanitary engineering systems proposed were effective against the causes of disease.

In the middle of the century Dr Snow investigated the spread of cholera and though he did not know the cause he established its source as contaminated water supplies. The need for clean water was recognised. The application of measures across the country was very hit-and-miss so public agitation, and it is rumoured the stench of the River Thames outside Westminster, eventually resulted in the setting up of public health authorities which could be properly controlled and had authority to enforce measures for the public good. The Public Health Act of 1875 lead to the provision of safe water and adequate removal of sewage and waste water from towns, see Fig 4.3(b).

4.3 MEASURING WATER QUALITY

Three factors are used to measure water quality: the organic material present, the turbidity and the numbers of certain microorganisms. The amount of organic material carried is measured indirectly. A water sample is taken and the oxygen content measured. The sealed sample is kept for five days at a standard temperature and then the oxygen content measured again. During that time the oxygen content will have fallen because of oxidation of the organic material present. The amount of oxygen used up is called the **biochemical oxygen demand** or **BOD**, and is calculated in milligrams per litre. Generally the higher the oxygen consumption the more organic material is present.

Turbidity is due to colouring matter in the water and suspended particles. The particles are filtered out, and weighed as milligrams of **suspended solids** per litre.

Monitoring harmful organisms is far more difficult. The pathogenic organisms from a few people suffering from a disease in a population are rare in the huge volume of bacteria-laden effluent a community produces. Finding these organisms is therefore extremely difficult. However untreated sewage carries large numbers of relatively harmless organisms called **faecal coliforms** (the *E. coli*-like bacteria that live in the guts of animals, including humans) which can be detected more easily. These organisms are called **indicator organisms**, because their presence indicates that there is untreated sewage in the water, possibly carrying pathogenic organisms. Their numbers are regularly monitored. There are limits, shown in Table 4.1, to levels of BOD, suspended solids and faecal coliforms that are allowed in discharge water.

Table 4.1
The British recommendations for effluent are:

BOD
less than 25 mg O_2 per dm^3
Faecal coliforms
less than 5000 cells per 100 cm^3
Suspended solids
less than 30 mg per dm^3

Some water-borne diseases are caused by viruses, which are difficult to detect. The viruses have to be concentrated on a membrane filter and then grown in cell culture. They are then detected using antibodies. The process takes days and is only really effective if there are large amounts of virus present. A new development is to use a gene probe. This is a piece of DNA or RNA which corresponds to sequences in the water-borne virus nucleic acid which will bind to the viral nucleic acid if it is present. The test can give results in less than one day and should be much cheaper than conventional tests when it is available. Potentially, probes could be made to detect a large number of different viruses.

QUESTIONS

4.5 List three factors used to measure water quality.

4.6 The table below gives readings for the BOD and suspended solids in the effluent from four factories and two farms:

Unit	Volume of effluent per week (1000 dm^3)	Average readings BOD	Suspended solids (mg per dm^3)
Factory A	18	28	55
Factory B	3	20	133
Factory C	11	115	30
Farm D	0.5	89	40
Factory E	6	18	28
Farm F	1	33	35

(a) Which of these units has effluent which exceeds the British recommendations?

(b) Each of these units produces a different volume of effluent. Calculate the total suspended solids for each one. Which contributes most to suspended solids pollution? Which causes the greatest BOD problem?

(c) Which of the factories could be a food factory and which a cement works?

(d) If one of the factories produces a hot water effluent, what effect will this have on decomposer microorganisms?

(e) Farm slurry is a very liquid mixture of animal faeces. Which of the farms could have slurry leaking into its drains?

4.4 HOW WATER IS PURIFIED

Providing safe water supplies requires a great deal of effort. The demand is huge but cannot be met everywhere. In Britain, water is supplied by Regional Water Authorities which are responsible for the supply of good quality water, maintaining rivers and supervising effluent treatment. For instance the Severn–Trent Water Authority covers over 8500 square miles in the Midlands supplying over 8 million people. These people use nearly 2000 megalitres of water each day, 60 per cent in homes, and the rest in industry. Industries use three times as much water again straight from rivers which may need treatment before it can be returned to the natural water system. The Authority also treats over 2500 megalitres of sewage per day from homes and industrial premises.

The supply

Water comes from three main sources.

Bore holes

These are very deep wells or shafts sunk into the underlying rock to draw water from the water table. Some rocks, called aquifers, are very porous and hold large reserves of water. This water may be good enough to go straight to the consumer, but in Britain it is chlorinated before it is distributed. Many water users have their own wells and bore holes.

Reservoirs

These also may have very high quality water but the water is usually treated to remove colour and suspended solids before it is distributed.

River water

This water is most likely to need treatment as river water may contain many undesirable materials. It will need filtration, clarification and chlorination before it can be distributed.

Water for domestic supply in the EC has to be treated so that it has no suspended material, no colour, its taste has to be acceptable and it must not smell unpleasant. There must be no harmful organisms and it should not carry too many salts which could deposit in pipework and use up large quantities of soap. Most purification techniques mimic the natural mineralisation processes which occur in soil and large bodies of water. Usually the processes are used in combination and chlorine is added as a disinfectant.

Purification

Filters

When water percolates through soil saprophytic organisms degrade organic matter in the water and hence mineralising it, and suspended solids are trapped as water filters through the soil particles. Filtration through synthetic filters mimics this process. The filter is a bed of carefully selected sizes of sand particles on gravel. Water is passed through and suspended solids are filtered out. In slow filters organic material may be degraded by saprophytic organisms around the filter bed particles. Any harmful organisms are likely to die as the temperature is low and they cannot compete with soil saprophytes. High quality water drains out from the bottom of the filter. Fig 4.5 shows the sort of portable filter people had to rely on before there were piped supplies of clean water.

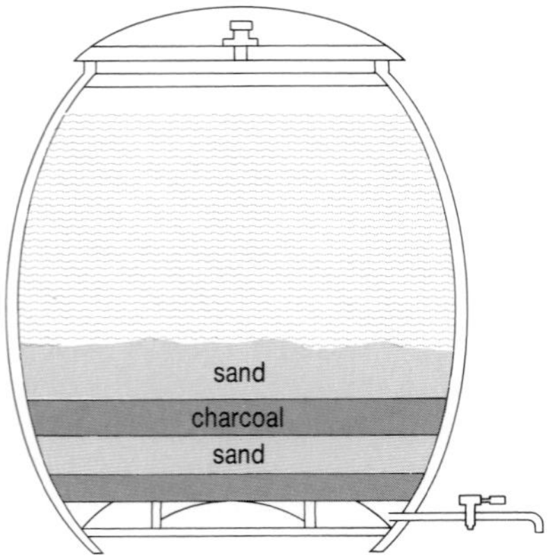

sand
charcoal
sand

".....to make a filter with a wine barrel, procure a piece of fine brass wire cloth of a size sufficient to make a partition across the barrel. Support this wire cloth with a coarser wire cloth under it and also a light frame of oak, to keep the wire cloth from sagging. Fill in upon the wire cloth about three inches in depth of clear sharp sand, then two inches of charcoal broken finely, but no dust. Then on the charcoal four inches of clear sharp sand. Fill up the barrel with water and draw from the bottom....."

Fig 4.5 A portable water filter used by settlers to provide drinking water in the USA at the end of last century. (Scientific American Cyclopedia of Formulas, 1903).

Reservoirs

Particulate matter in the water of large lakes settles to the bottom – even small particles such as clay eventually settle out. Organic matter is oxidised rapidly by micro-organisms in the nutrient poor water, temperatures are quite low and, except in warm climates, potentially harmful micro-organisms die. In a reservoir the same thing happens. Also colouring matter in the water, such as tannins, is gradually bleached out and hydrogen carbonates are converted to carbonates which settle.

Purification plants

Some water contains so many impurities that it needs more thorough purification than either of the two processes already mentioned. It is passed through mesh screens to remove debris and may need to be aerated to remove smelly gases. The water is treated with chemicals to coagulate suspended solids making a **floc**. Some colouring matter and bacteria are trapped by the floc which is drained off to a sludge lagoon. Water hardness is adjusted by precipitating salts or by ion exchange methods, and chlorine is added to kill any remaining micro-organisms. The water is then filtered through a filter bed to remove any remaining particles. Acid water corrodes pipework and alkaline water leaves deposits, so the pH is adjusted to close to neutral. More chlorine is added to ensure a residual chlorine level which reduces the growth of microorganisms in the pipelines and acts as a safeguard against minor leaks into the water supply.

In areas of high demand rapid gravity filters are used, as shown in Fig 4.6. These can deal with large volumes but are not as good at removing tastes, colours or bacteria as ordinary slow filters. Air is blown through the filters every few days to clean them as they are easily clogged.

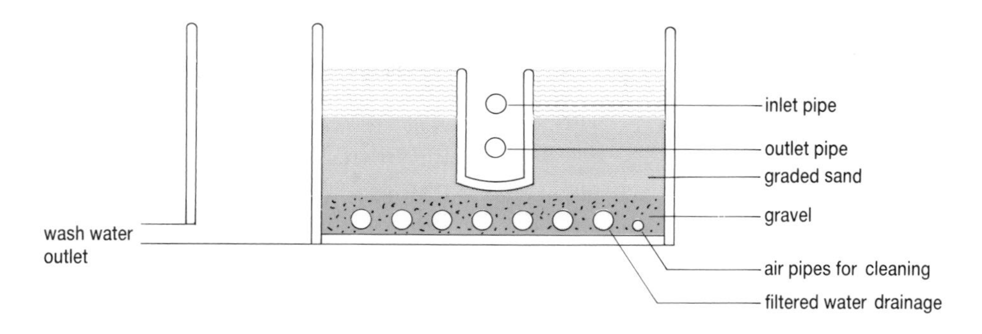

Fig 4.6 A rapid gravity filter bed.

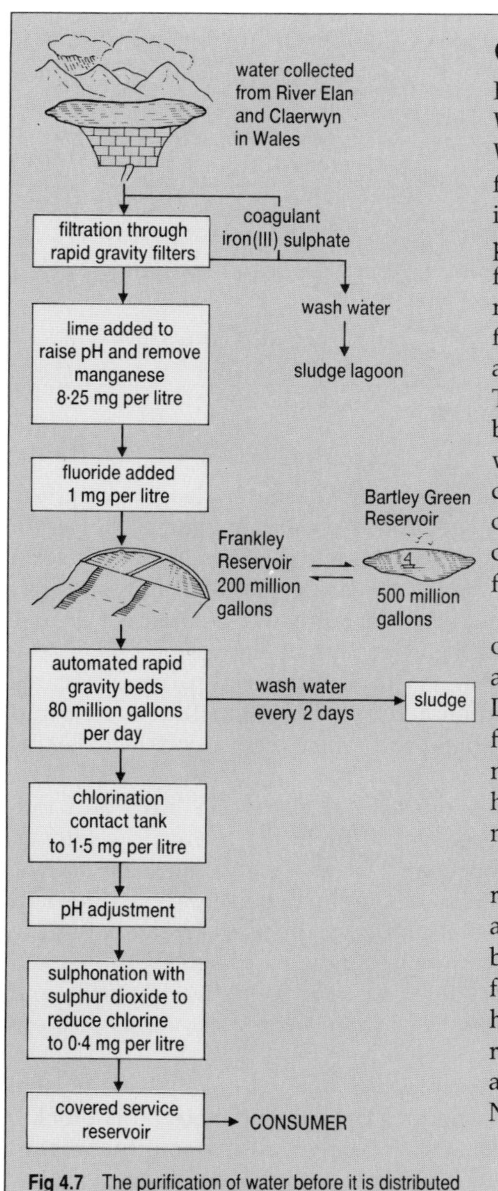

Fig 4.7 The purification of water before it is distributed in Birmingham.

Case study: Birmingham's water supply

Fig 4.7 illustrates the processes undergone by water taken from central Wales to provide drinking water in Birmingham by the Severn–Trent Water Authority. Up to 80 million gallons of water a day can be taken from the Elan Valley in central Wales to supply Birmingham. The water is acidic, pH 5, because it drains through moorlands but there is no problem with nitrates as local land uses are mainly sheep farming and forestry, and neither of these needs the addition of large amounts of nitrate fertiliser to the soil. The water is held in reservoirs, then filtered, fluoridated and some chlorine added before it flows through an aquaduct to Birmingham where it is held in reservoirs at Frankley Treatment Works. The water is already drinkable when it leaves Wales but at Frankley the water is treated to produce high quality drinking water. The water is very soft so its pH is adjusted to about pH 9 before distribution. A final dose of chlorine is added to protect against contamination in the pipeline. The demand from the 1.2 million consumers is about 70 million gallons a day, so there is a back-up supply from the River Severn if necessary.

Seagulls are a problem when they move inland to overwinter on the outskirts of the city. During the day they scavenge in rubbish dumps and other places then roost on the open reservoir waters at night. During the winter months the water becomes contaminated with gull faeces and enteric organisms, raising the bacterial count from the normal very low levels to extremely high numbers. As a result the water has to be more strongly chlorinated than at other plants and efforts are made to dislodge the gulls from their winter habitat.

The distribution network serves five zones each with covered service reservoirs; once the water has been purified it does not see day light again until it is in someone's house. The whole system was engineered by the original City of Birmingham Waterworks Company, and gravity-fed pipelines are used to distribute the water. There is a fall in land height between the Welsh reservoir and Frankley, and the water supply regions within the city are arranged by altitude. The only water which is actively pumped is that going to the local Frankley service reservoir at Northfield and the wash water used to clean the filter beds.

The future

There are occasional problems in water purification with the levels of ions such as manganese in the water. From 1992 onwards lower levels than are currently acceptable will be enforced within the EC, so research is being done on ways to reduce them. Clarifiers working on the sludge blanket principle, in-line filtration and dissolved oxygen flotation have all been investigated. In-line filtration is successful at removing the ions but the filters need frequent cleaning and it is not practicable to have them out of service so often. The sludge blanket technique involves treating the impurities in the water with a polyelectrolyte coagulant before the water is pumped up from the bottom of a large hopper. The flocculate formed collects as a sludge blanket just below the water surface. Clean water is drained off from the surface and sludge can be siphoned off to a sludge lagoon and dried.

The most likely process to be used is that of **dissolved oxygen flotation**. A polyelectrolyte coagulant, which is mainly aluminium sulphate, is added to the water which forms a complex. This flocculates minerals, colour and fine particles as the water passes through a slow-stirring chamber. Aeration nozzles then bubble air through the water and lift the flocculate up to the surface where it can be skimmed off before the water is chlorinated and distributed.

Water can be purified by filtration, by standing or by a combination of treatments with chlorination.

QUESTIONS

4.7 What are the main sources of water for drinking?

4.8 Why does allowing water to stand in reservoirs result in a fall in the numbers of micro-organisms present?

4.9 Explain how filters can be used to obtain pure water supplies.

4.5 WHAT'S IN SEWAGE?

Sewage is what is in sewers – a complicated mix of materials from a variety of sources. Each person in Britain generates an average of 500 litres of sewage a day; about a third from domestic use, the rest from industrial uses. The largest water user is the electricity generating industry but chemical industries and engineering also use large amounts. Street drains, called **storm water drains**, take, among other things, rainwater, bus tickets, leaves, grit and car keys. The drains from **industrial premises** carry rainwater but also materials from their activities. These include metals and inorganic ions, oils, greases, solvents, food waste from food factories as well as a variety of other materials.

If you live in Britain you are responsible for about 170 litres of a rich soup of effluent from *your home* each day. The 30 litres from the kitchen carries a range of detergents, disinfectants, bleaches and scouring agents. There may be other items too – tea leaves, spaghetti, soil, spiders, dyes and cigarette ends, to name just a few that go down the sink. In the bathroom you use an average of 35 litres of bath water carrying grease, bath oils and soap. Each flush of the toilet uses 9 litres to carry away faeces, urine and large amounts of toilet paper. Even with all these additions over 99 per cent of sewage is water.

There are large quantities of grit, minerals and so on, but about two-thirds of the material in sewage is organic. The organic material is quickly broken up and is carried in suspension. There are also micro-organisms, some of which cause disease. Water treatment must remove the organic materials, kill the harmful organisms and reduce toxic chemicals to very low levels.

4.6 USING MICRO-ORGANISMS TO CLEAN UP WATER

Bacteria and fungi secrete enzymes into the environment which break down materials, and protozoa can take in particles to degrade them intracellularly. Though no organism can break down everything, between them microorganisms can degrade all naturally occurring carbon compounds and quite a few synthetic ones.

A sewage treatment plant harnesses natural decay processes and provides the best possible conditions for the growth of micro-organisms and hence the efficient breakdown of organic material. Some of the degradations are anaerobic but most require a high oxygen concentration so effluent treatment plants have to be engineered to fulfil this requirement. The final breakdown products are carbon dioxide, which is released to the atmosphere, nitrates, sulphates and phosphates, which remain as dissolved ions in water and are discharged. A large amount of sediment and sludge accumulates which can be decomposed anaerobically to produce useful products such as methane.

Organic material in effluent is degraded by the enzymic action of a variety of micro-organisms.

What happens in a sewage works?

In a modern sewage works the effluent undergoes a number of processes to rid it of as much debris, organic matter and dissolved chemicals as is reasonably possible. In the process harmful organisms die. The main stages are outlined in Fig 4.8.

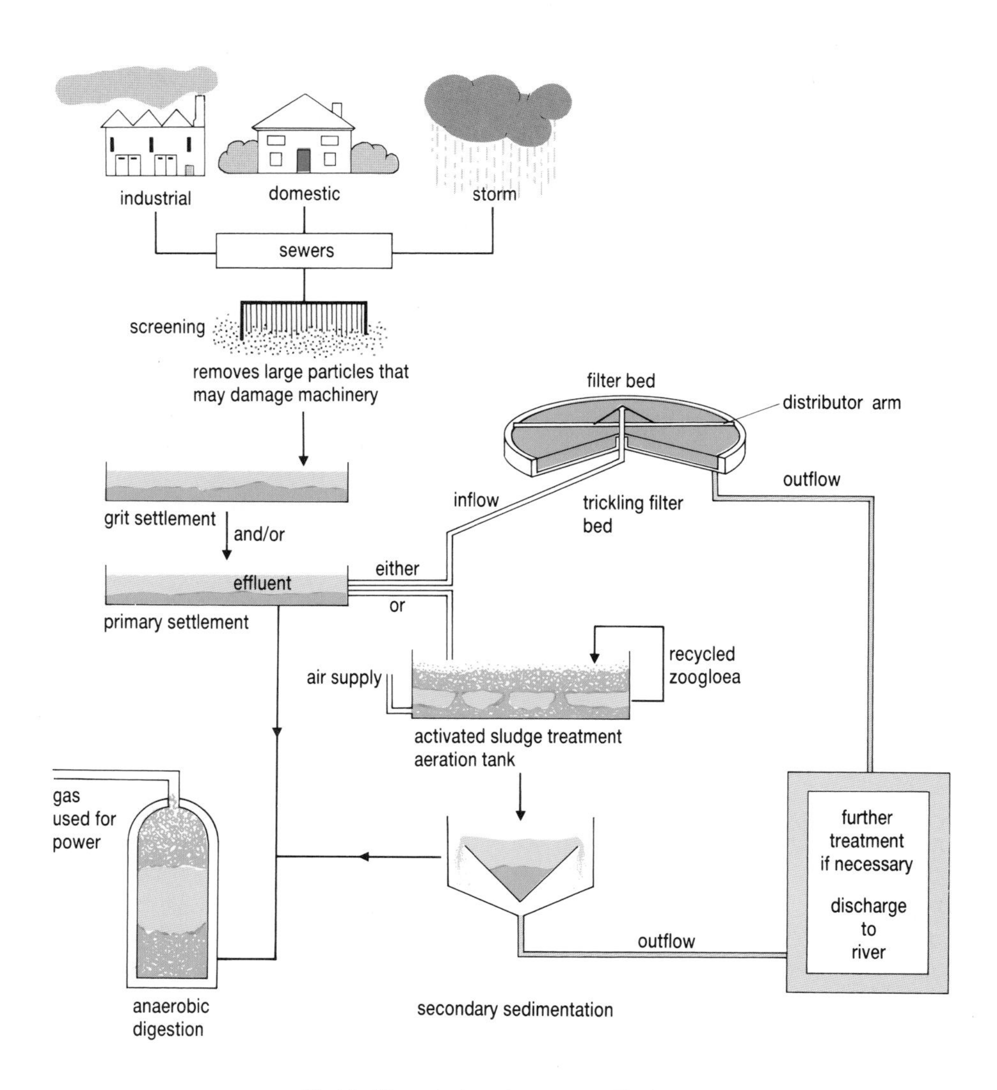

Fig 4.8 The main stages in sewage purification.

Entry

Sewage is screened and then enters a grit settlement tank for a few hours. Grit and stones settle out and some organic materials flocculate and settle. The sediment is called primary sludge and needs further treatment. In the grit tanks there is some microbial breakdown of organic matter and some of the micro-organisms in sewage settle out, but most purification happens in the next stage. After grit settlement the water can pass through a trickling filter or it may enter an activated sludge treatment tank.

Trickling filters

The trickling filter bed is made of carefully graded stones, grit and clinker and a complex ecosystem is allowed to develop in it. Bacteria, fungi and protozoa degrade organic matter as the effluent trickles through the bed, and these are eaten in turn by other organisms such as predatory protozoa and insect larvae. Different organisms with different kinds of metabolic activity dominate in different parts of the bed. By the time water has passed through to the drains at the bottom, most matter has been mineralised.

Fig 4.9 An aerial view of a sewage treatment works. Both circular and rectangular filter beds can be seen; the smaller circles are sedimentation tanks.

Activated sludge

The activated sludge treatment uses a starter culture of a mixture of organisms called the **zoogloea**, including aerobic bacteria and protozoa. Some of the organisms found in a zoogloea are listed in Table 4.2. One species, *Zoogloea ramigera*, secretes a gum that flocculates particles together and colonies of organisms work within the floc breaking it down. These organisms need a high oxygen concentration which is supplied by stirring or by bubbling oxygen through the tank. After a few hours in the tank the zoogloea reduce the BOD by about 90 per cent. A portion of the zoogloea is then syphoned off when the water leaves the tank to inoculate the next batch of sewage. There are variations on this basic design: one example is a cylinder coated with a thin film of organisms on the outside that rotates through a trough of effluent.

Table 4.2 Organisms in zoogloea

Organisms in zoogloea
Zoogloea ramigera
Beggiatoa alba
Sphaerotilis
Pseudomonas
Alcaligenes

The zoogloea is sensitive to certain contaminants, for example if there has been accidental sewage contamination with heavy metal ions the effluent may have to be held back and diluted with ordinary sewage to avoid killing the zoogloea. Another problem comes from foaming detergents which reduce the movement of oxygen from air into the water and trap organisms in the foam away from their food source.

Digestion

After these treatments the water has few organic contaminants or suspended solids left. Any remaining particles or zoogloea are settled out in a secondary sedimentation tank. Sediment from the primary and secondary sedimentation tanks is treated before disposal. Sometimes it is simply dried and burnt, but this is a waste of a potentially valuable resource. Often the sediments are put in large, warm tanks called **digesters** where anaerobic breakdown can take place. The sediments are broken down by organisms such as *Clostridia* to acetate and other reduced carbon compounds and hydrogen. These are used by methanogenic bacteria producing methane which can be used as a power source.

After digestion, some solid sediment remains which is dried and used as landfill or is dumped at sea. Very few harmful organisms survive the treatments so it is usually safe to use the sediment as fertilisers, though in many industrial areas the metal content may be high so only a limited amount can be spread on agricultural land.

In The Netherlands another process has been developed to purify waste water entirely anaerobically. Anaerobic bacteria are immobilised onto sand particles in a large digester. The waste water is fed in and the organisms metabolise materials in the water. A useful by-product is methane.

Other sewage treatment systems

In many places the population is too scattered for a central sewage works to be economical, and in others water is scarce so it must be used sparingly. Alternative sewage treatments have been used in these situations, some more successfully than others. The most common method is the use of pit latrines which are be designed to discourage the entry of flies and are concrete-lined to prevent the contents from contaminating ground water supplies. Within the pit there is microbial breakdown of faecal material.

Digesters

The organic material in sewage is too valuable a resource in many places to be used to feed bacteria so there are systems which put the organic matter to better use. Households can use sewage, along with food waste and animal manure, in digesters to generate methane for cooking and power; more about this can be found in section 4.8. The high temperatures generated within the digester are enough to kill most pathogenic organisms. A number of different digesters and 'waterless' composting toilets have been designed, though few of the latter are as yet in wide commercial production.

Sewage lagoons

Sewage is pumped into large, open ponds and the natural decay processes degrade organic material. The lagoons are shallow enough to ensure that there is enough oxygen diffusing in for aerobic breakdown. However breakdown is slow and in warm climates some harmful organisms may survive in the water.

Crop irrigation

Domestic water and sewage can be used to irrigate crops. Higher numbers of organisms are acceptable in water used for crops such as trees and cotton than water intended for consumption, but ideally irrigation water should not carry any harmful organisms at all. One method currently on trial in Egypt is to use the water on reed beds before using it to irrigate food crops. This technique was used in Britain during the last century but it was not very efficient. The modified method is to run sewage into a settlement tank to make a sediment which can be used or burned; the water is then run into impermeable shallow channels. These are filled with gravel and, as the channels are built on a slight slope, the water runs down through the gravel. Reeds planted in the gravel ensure aerobic conditions and micro-organisms can break down organic material. Reed roots take up nitrates produced by bacterial action. The reeds can be harvested as a separate crop and the water from the channels, which should now have very few harmful organisms in it, can be used to irrigate crops.

QUESTIONS

4.10 What is a floc and what is its role in sewage treatment?

4.11 List the major physical, microbiological and chemical changes in effluent as it passes through the treatments in a sewage works.

4.12 What are the main advantages of using micro-organisms to purify water?

4.13 What useful products are made by anaerobic digestion of sewage sludge?

4.14 **(a)** Suggest the possible sources of pollutant materials that occur in river water.

The table below shows the results of a chemical and microbiological analysis of samples of two different river waters (freshwater), both being suspected of recent pollution.

River sample	BOD (mg per dm^3)	NO_3^- (mg per dm^3)	NH_4^+ (mg per dm^3)	coliform content (*E. coli* per 100 cm^3)
A	32	60	0.75	> 18 000
B	3	4	0.10	< 50

(b) (i) Explain briefly the value of each of the four tests used as indicators of water pollution.

(ii) Suggest (with reasons) whether either of the river samples A and B would be suitable for domestic or industrial purposes.

(c) Outline any three suitable methods which may be used to produce large-scale purification of local water supplies.

(JMB)

4.7 CLEANING UP THE ENVIRONMENT

There are many waste materials causing ecological and aesthetic problems in our environment. Microbial breakdown of these has been investigated because of economic problems such as oil washing up on tourist beaches, or health problems arising from pollutants. Pollution is usually treated mechanically or chemically, using filters or carrying out chemical cleaning processes. Now, as the energy required for these processes gets more expensive, it is profitable for people to turn their attention to the microscopic organisms that do the job for free.

Oils and plastics

Hydrocarbons from crude oil are made into a variety of oils, polymers and plastics such as PVC, polyethylene and polyurethane. They cause a variety of disposal and pollution problems, particularly waste engine lubricating oil and plastic wrappers. They are either not degraded at all or are degraded very slowly by soil organisms. However, specialised groups of micro-organisms can grow in fuel oils and lubricants if there is some moisture available. Crude oil in oil fields harbours many different micro-organisms but there is not enough water for their growth. Once moisture enters the oil after pumping up, or in engines and machinery, the micro-organisms will grow.

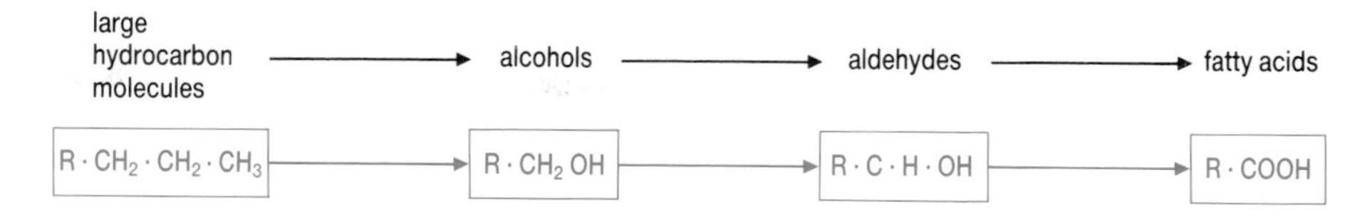

Fig 4.10 The breakdown of hydrocarbons by micro-organisms.

As they grow, the micro-organisms break down hydrocarbons generating corrosive metabolites such as organic acids, sulphuric and nitric acid and ammonia. Fig 4.10 illustrates the general pathway. If any oxygen is present, electrochemical reactions take place when acids come into contact with steel, causing corrosion; hydrogen is also generated. Some organisms produce hydrogen sulphide which can cause explosions in fuel lines. These processes can cause a range problems such as the rusting and sinking of coal carriers, blocked aircraft fuel lines and corrosion, as shown in Fig 4.11(a). Other materials are degraded too: micro-organisms such as Pseudomonads degrade water-based paints and polyurethane coatings, corrosion inhibitors in car radiators, adhesives and synthetic materials.

These problems have led to the investigation of hydrocarbon-degrading organisms, and as a result it was realised that some of these specialised species could be used to degrade unwanted oils and plastics. Members of over 20 different groups of bacteria can degrade hydrocarbons, so there are many potentially useful organisms. *Pseudomonas aeruginosa*, for example, degrades plasticised vinyl quite quickly, and a fungus, *Cladosporium resinae* degrades paraffin-based fuels such as aircraft fuel. Many organisms can degrade plasticisers, usually organic acid esters, so any plastic containing plasticisers are usually degradable. Some companies now deliberately use packaging plastics which are microbially degradable to reduce pollution – see Fig 4.11(b).

A genetically engineered strain of *Pseudomonas* has been found to be efficient at cleaning up oil spillages and is now in commercial use, and more investigations into the use of micro-organisms to clean up marine pollution are being started.

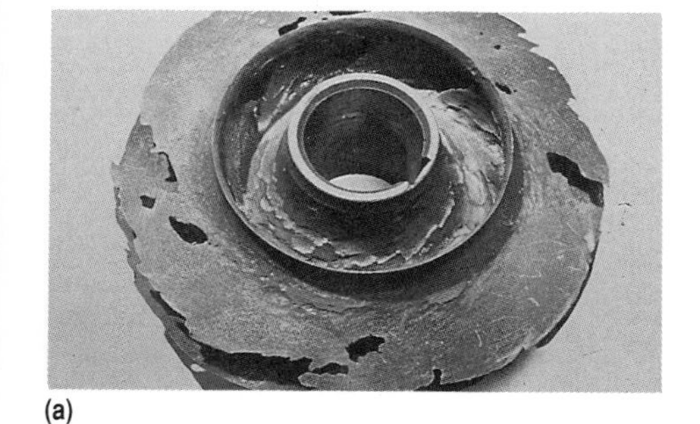

(a)

(b)

Fig 4.11 **(a)** The activity of micro-organisms such as *Thiobacillus* make mine drainage waters very acidic. Any metal items items in contact with the acid waters corrode very rapidly. This piece of pumping machinery had a working life of only six months before it had to be replaced.
(b) Some manufacturers and retailers have responded to public concern about pollution by using biodegradable packing for their products. This label can be found on some carrier bags.

Reclamation of industrial sites

Many industrial activities, particularly mineral extraction, leave spoil tips or lagoons with large amounts of waste materials. Reclaiming these areas is difficult as the physical environment is often unsuitable for the growth of plants and animals. Spoil heaps are unstable and short of plant nutrients, particularly nitrates. For many years now symbiotic nitrogen-fixing bacteria have been inoculated into reclaimed spoil heaps and other areas at the same time as clover and grass seed. The bacteria in clover root nodules encourage growth which stabilises the ground, and other species are able to establish. As the clover decays, nitrate is made available to the other species, thus enhancing their growth, and a natural succession of plants and the animals that feed on them is set into motion. The reclamation of industrial areas is shown in Fig 4.12. Two examples of reclamation works currently in progress illustrate the ways in which micro-organisms can be used.

Fig 4.12 The spoil tip at this Nottinghamshire Colliery is being reclaimed and turned over to agriculture. The spoil heap needs stabilising and seeding to encourage the growth of tolerant pasture species.

Example 1

Lagoons from sand and gravel works can be problems. One large company had excavation pits which, when finished with, filled with water. The lakes formed were unsuitable for much life as the water had a very low pH and contained no organic sediment to act as a source of nutrients. To reclaim the land, the company added a large amount of activated sludge which acted as a source of decomposer organisms to replace the existing acid-tolerant chemoautotrophs, and as a source of organic material. Later hydrated lime was added to adjust the pH to closer to neutral. The sewage sludge acted as a primer and allowed the establishment of plants and other organisms and the eventual development of a stable ecosystem.

Example 2

Another company is using micro-organisms to reclaim an old gasworks site heavily contaminated with a range of organic molecules such as oil tars and cyanides. The micro-organisms were isolated from the site, which had effectively become a giant selective medium for those organisms which could tolerate and metabolise the contaminants. These were isolated from the soil, grown in large quantities in the laboratories, sprayed over the contaminated soil, and are given occasional nutrient sprays.

Chemical contamination of the environment

There are a number of chemicals entering the environment whose concentrations cause concern. Sulphur dioxide, together with nitrogen oxides, makes rain acidic causing environmental damage. Metal ions such as mercury and cadmium enter soils and water as a result of many industrial processes, from paper making to 'metal bashing' industries. Heavy metals accumulate in food chains and are toxic in quite small amounts.

There is growing concern about the amount of nitrate from agricultural fertilisers and effluent treatment plants entering water. Excessive amounts of nutrients such as nitrates and phosphates lead to eutrophication of natural waters and in some cases human illness. Worries about a rare condition, methaemoglobinaemia, caused when excessive amounts of nitrate are taken in by young children, has led to the imposition of limits to the amount of nitrate allowed in drinking water in the US and by the EC.

Reduction of these pollutants is a difficult task and often expensive when done chemically, so alternative, biological, ways to reduce pollution hazards are being sought.

Sulphur dioxide

Sulphur dioxide in the atmosphere comes from a variety of sources but coal- and oil-burning in power stations are major sources. Coal and oil are fossilised plant material containing sulphur compounds. When coal or oil is burnt these are oxidised to sulphur dioxide which is released to the atmosphere. Sulphur dioxide combines with moisture in the atmosphere to make acids that contribute to the acid rain problem. Fig 4.13 shows some of the spread of acid rain over Northern Europe.

Power station coal produces over 2 million tonnes of sulphur dioxide in Britain each year. Removing the sulphur dioxide from effluent gases can be done chemically, but techniques are being developed which will remove sulphur from the coal before it is burnt. Current ideas include using *Thiobacillus* to metabolise sulphur-containing pyrites in the coal but the process is not practical or economic on a large scale at the moment.

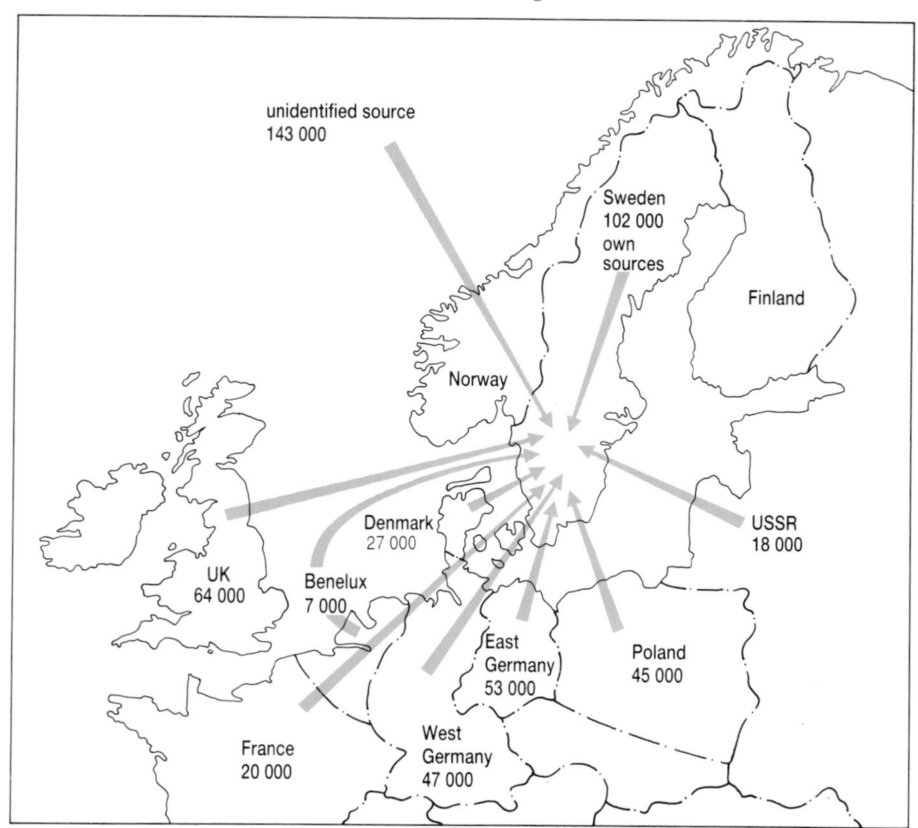

Fig 4.13 The spread of acid rain from north Europe to Scandinavia (tonnes sulphur pollutants).

Metals

The extraction of metals from low concentrations in water is both physically and chemically difficult. Often it is not worthwhile, except for valuable metals such as gold. Increasing awareness of heavy metal pollution, however, has led to pressure for better techniques to extract metals from low concentrations in water. Some organisms such as the unicellular alga *Scenedesmus* and some fungi take up metal ions from dilute solutions for their own metabolism. They can accumulate these against a concentration gradient in non-toxic forms. Mycorrhizal fungi (those that live in a mutualistic relationship with various plants) have been shown to take up metal ions from dried sewage sludge that has been used as a soil conditioner. Some bacteria can even accumulate mercury without coming to too much harm.

Fig 4.14 Helicopters are used to spray fertiliser at a rate of 150 kg per hectare on land which is difficult to reach by more conventional means. Much of this fertiliser is likely to end up in ground water.

Some research has been done using algae to extract precious metals such as gold from seawater and jewellery industry wastes, where the concentration may be too low for conventional methods. The development of photobioreactors, where algae are grown in controlled conditions, could make this more feasible. An alternative approach is to grow organisms which bring about physical changes in waste waters which make polluting metals precipitate in forms which are easier to extract.

Nitrates

Excess nitrates in water are very difficult to remove chemically, and are proving difficult to control. One approach is to reduce excess nitrate at source. Much comes from agricultural fertiliser (see Fig 4.14), a high proportion of which leaches into watercourses. Either farmers have to be persuaded to use less fertiliser more effectively, or alternative forms of nitrate which are less soluble could be developed. Another approach is to engineer bacterial nitrogen-fixing genes into crop plants to reduce the need for fertilisers, but much work remains to be done in this area.

Potentially, organisms can be used to take up nitrate as a nutrient from polluted water. Several systems are being investigated. In one, contaminated water is passed through an ion exchange column which exchanges hydrogen carbonate ions for the nitrates. The nitrates are then washed out of the column into another where denitrifying bacteria reduce the nitrate to nitrogen gas which is released to the atmosphere.

QUESTION

4.15 You are provided with a fungus, which is known to degrade some plastics that are used to make carrier bags, and two carrier bags. One carrier is from The Whole Earth Muffin Company, a company that specialises in natural foods and uses biodegradable plastic bags, the other is from Monstaveg, a greengrocer. Outline how you would investigate the biodegradability of the two carrier bags by your organism at soil temperatures.

4.8 TURNING WASTE INTO FUEL

Some sewage plants use sediments to generate methane fuel. There are other methods too in which a waste from a process can be used to make fuel. Energy from biomass, in the form of burning dried animal dung and wood, has been around for thousands of years, but these are becoming scarcer. In most countries fuels, particularly oil, are costly or scarce, or have to be imported from other countries.

Oil has to be paid for out of a country's foreign exchange which is earned by making or growing a product which can be sold abroad. Many of the poorer countries of the world find themselves caught in a spiral: they need fuel, so they have to focus their economy on earning foreign exchange instead of being able to concentrate on the needs of their own population. For example cash crops such as coffee, which can be exported, may be grown on good agricultural land while the local people are malnourished because of inadequate food. The money from exports is used to pay for imported oil instead of for developing the economy, improving health care, providing safe water and schooling. These activities are paid for by money borrowed from other countries, and as more foreign exchange is needed to repay the loans and interest payments, so more cash crops are grown, and so on.

Escaping this spiral is a powerful incentive to find ways to generate fuel from local materials, and, for a scattered population, on a small scale. There are two major programmes underway: generating alcohol as a fuel from biological materials and the use of biogas digesters as local energy sources.

Fig 4.15 Many cars can be converted to run on petrol diluted with ethanol; but may be worthwhile for a car manufacturer to produce an engine which runs on pure ethanol in countries where there is a large market for alcohol-powered cars.

Alcohol

Ethanol and other alcohols are important raw materials in the manufacture of plastics and other synthetic materials. Ethanol can also substitute for mineral oils and is used as a transport fuel (see Fig 4.15). Petrol mixed with ethanol, or **gasohol**, has a long history as a fuel for cars – there were alcohol-driven cars in the 1930s – but in many countries petrol is cheaper than ethanol, so there is no profit in using ethanol as fuel. However, if a country has to import oil to make petrol, it may be worthwhile using ethanol, if it can be made cheaply enough. Obviously the cost of ethanol manufacture is the crucial factor.

In most industrialised countries ethanol is made from oil, but alcohols can also be made by microbiological methods. During World War I butanol and acetone for explosives were made by *Clostridium acetobutylicum* fermenting starches and molasses anaerobically. Unfortunately it became cheaper to make them from oil so microbial production stopped decades ago. The only alcohol made microbially now is that in drinks such as wine and whisky.

If petrol is very expensive, microbial ethanol production may be economic, as long as power for the factory and the raw materials are cheap. Some microbial ethanol is made in the USA as an octane enhancer for use in states where lead cannot be added to petrol. The process is not really economic as there is neither a cheap fermentable substrate nor cheap power for distilleries. The costs of setting up a factory from scratch are very high and many commercial companies cannot wait long enough for the factory to start operating profitably. The case study at the end of the chapter illustrates the gasohol programme in two different countries.

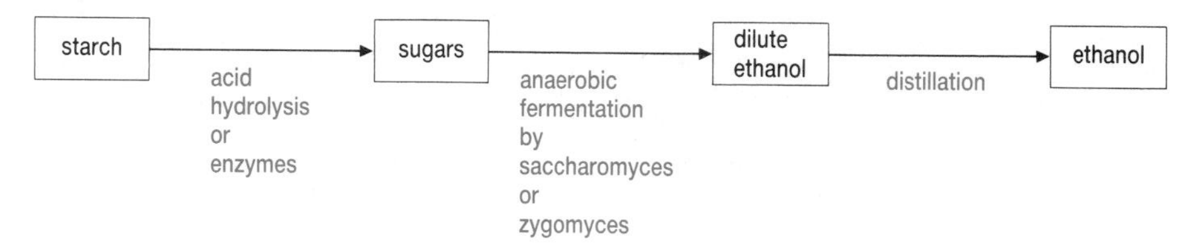

Fig 4.16 The breakdown of starch to ethanol.

The basic fermentation process is shown in Fig 4.16. Many wastes are used as raw materials for ethanol production. There are agricultural wastes such as straw and maize cobs and processing wastes such as fruit and vegetable peelings, leaves, pods and husks. Other industries including the wood, paper and cotton industries produce pulp liquors, sawdust and fibres which also contain waste carbohydrates. These wastes cost money to transport to a dump or to process before discharging into water; in many places they are simply dumped into rivers causing gross pollution, but they can be used for ethanol production.

Yeast cannot metabolise waste carbohydrates directly; they have to be broken down to simple sugars, or **saccharified**, first. This can be by acid hydrolysis but this adds to the cost of the process. Enzymes are now available to do the job quickly and leave no acid residues. When fermentation is finished there is about 9 per cent ethanol in the solution which has to be distilled off to be useful. The cost of power for the distillation may be the deciding factor in the economics of the process as a whole. If a fuel with a high energy value has to be used to generate a fuel with a low energy value the project is a waste of time, unless there are other socio-political factors which outweigh cost.

Fig 4.17 Rubbish has to be treated when it is dumped by compressing, compacting or baling. Landfill sites are filled quickly and it is becoming harder to find new ones. Some of this rubbish could be used to generate power.

Biogas digesters

Biogas digesters are potentially useful in both developed and developing countries. In industrialised countries the amount of refuse is increasing, much of it organic. Materials such as cardboard are awkward to dispose of except by burning; baling it and dumping is expensive (see Fig 4.17). The main advantage of using waste to generate methane is that it solves the waste disposal problem; generating power this way is not usually cheaper than other methods in industrialised countries.

In developing countries fuel may be expensive, particularly in deforested areas where fuel wood is scarce, and there may also be a sewage disposal problem. Biogas digesters, using wastes to generate methane are valuable, but expensive, resources in these places. The gas generated is used for cooking, lighting, as diesel substitute and for powering refrigerators. The sludge left at the end is a useful fertiliser. As well as the immediate benefits of generating fuel this way, deforestation and soil erosion can be reduced and the soil fertility increased. It is also more efficient: a fire burning dried dung is only one-sixth as efficient as an energy source as the same dung used in a fermenter. Many pilot projects are underway in various parts of the world and in India and China biogas digesters are in regular use, China having several million digesters installed in villages.

Fig 4.18 A domestic-scale biogas fermenter on trial in India. A range of organic waste can be fermented to produce a mixture of methane and carbon dioxide which is held in a reserve until needed.

The process

There are many designs used, from simple plastic bags laid in trenches to concrete-lined pits with input hatches and outflow vents and regulators (see Fig 4.18). The gas does not flow evenly so a storage device with a regulator valve is needed to stop the biogas cooker going out half way through cooking a meal. Better designs are being investigated in which the fermenter contents are stirred or the bacteria held on a fixed film. However, many of the people who need methane the most cannot afford high technology fermenters, nor can they run and maintain them, so efforts are also being made to produce appropriate designs for poor people.

Most materials can be digested including human and animal excreta, vegetable peelings and animal wastes. The digesters need to be kept warm, at least 15°C, for fermentation to take place, and the interior of the digester must be oxygen-free for efficient digestion.

Anaerobic bacteria first degrade organic matter to methanol, hydrogen, formate and acetate, then the methanogenic bacteria use these to generate methane and carbon dioxide, a mixture known as **biogas**. Maintaining the digester is difficult as the methanogenic bacteria are sensitive to many factors in the refuse such as detergents, high fatty acid concentrations and heavy metal ions. Sulphides may make trace elements unavailable to the bacteria and stop their activity.

QUESTIONS

4.16 What is gasohol and what is it used for?

4.17 How could biogas contribute to the household economy of a family in rural Nepal?

4.18 List four materials that could be used as starter materials for the production of biogas and gasohol.

4.19 Why don't people living in Great Britain use gasohol?

4.20 Outline the fermentations used to make biogas and gasohol.

Case study : gasohol

Gasohol, a mixture of 80 per cent petrol with 20 per cent dehydrated ethanol, is often quoted as the answer to soaring petrol import bills, but this is not always the case. Developments like the gasohol programme have to be very carefully thought out if they are to benefit a country's economy and not handicap it. In Brazil the programme has had some success, but a similar attempt in Kenya failed for reasons which were not biological.

Brazil

Brazil has oil fields but can only supply part of its needs. The gasohol programme was started in 1975 and had success within the first few years. The target was to save 40 per cent of petrol consumption, about 10 billion litres, by 1985. By 1982 over 6 billion litres of gasohol were being produced. As ethanol was substituted, petrol imports were held level in the face of increasing demand, so that petrol imports and interest payments were held roughly equivalent to the foreign earnings of about US $20 billion.

The petrol pumps and cars already in use could cope with the new fuel if they had minor modifications which were quite cheap to carry out. The new fuel was sold at half the price of pure petrol as an incentive to people to switch, and a local programme was started to make cars which were designed to run on gasohol or pure ethanol. The market was large enough to make redesigning an engine for Brazil worthwhile. Cars with the special engines can run on pure ethanol, but it is usually blended with at least 3 per cent petrol to stop people drinking it! The ethanol content of gasohol has to be limited because it causes petrol pump and carburettor corrosion.

At the start the programme was only just viable, but by 1983 the world oil prices had risen enough to make the process economic, though ethanol production was still below target. The country's sugar mills were not fully used so they were used instead as distilleries to concentrate the fermented ethanol. The costs are less today because factories can be built using cheap local steel instead of expensive imported stainless steel.

Fig 4.19 In the cane crushing plant cane is passed through rollers to squeeze out sugary juice. A mass of fibrous stem material, called bagasse, is left which is dried and burnt in the distillery.

The crucial factors in the production of gasohol are the source of cheap fermentable carbohydrate and the power costs of the distillery. One of Brazil's main cash crops fulfils both these needs and solves a waste disposal problem at the same time. After harvesting, sugar cane is crushed by rollers and the juice extracted. This leaves the fibrous part of the cane as a waste called **bagasse** which can be dried and some of it used to make hardboard or filler. The juice is processed and sugar crystallises out. It is extracted two or three times, leaving a syrup containing fructose and glucose but it is uneconomic to try to extract more sucrose from it. The syrup is called molasses and is used for animal feed or other uses, or is dumped. Liquid molasses are fermented by *Saccharomyces cerevisiae* to produce impure dilute ethanol. The other waste, dried bagasse, is burnt to power the distillery to purify ethanol. Fig 4.19 shows sugar cane processing.

There are always advantages and disadvantages to any process. When Brazil diverted its molasses into alcohol production the world price of molasses for animal feed went up, so it was then worthwhile selling the molasses again and buying oil! It has been said that the anhydrous ethanol could be better used as a raw material in the chemical industry to make ethylene for plastics, thus reducing the import bill for petroleum derivatives.

There are also drawbacks to using sugar cane: it was calculated that it would need 15 per cent of the world's production to reach Brazil's 1985 target for gasohol production. Sugar cane takes a long time to grow; it is a seasonal crop and can only be harvested twice a year. If the programme is to be useful it needs a continuous supply of fermentable material together with cheap power for the distilleries. There are alternative crops such as manioc (cassava), which is a staple food, however it contains starch which

Fig 4.20 In Brazil, gasohol and anhydrous ethanol is sold at garages just like ordinary petrol. Gasohol was subsidised at first to encourage people to switch to the new fuel.

has to be converted to sugar before fermentation can take place. This begs the question of whether a farmer should be eating his manioc or using it to drive his tractor. Other problems have arisen with rain forest clearances and the displacement of small farmers from their land by larger growers. The programme would require half of the land currently cultivated to make Brazil self-sufficient. Nevertheless, the programme has resulted in a drop of 20 per cent in Brazil's oil consumption and a reduction in the import bill.

Kenya

Attempts were made to set up an ethanol production plant in Kisumu in Kenya in 1979. Molasses was available from a large sugar factory and from other factories in the area as the raw material for fermentation. It was hoped that when the plant was built it could also be a source of other chemicals such as citric acid, fertiliser and carbon dioxide. Waste from the processing was to be used to generate methane which would supply a large part of the factories' fuel needs. The factories would also provide work and stop pollution caused by dumping surplus molasses in the local rivers.

The project was beset by problems. New buildings were needed to house the new processes, but delays and the late arrival of promised money meant that costs soared to double the original estimates – over £50 million – and the contractors were bankrupted. The factory had to use outside power to start up, as its own power production could only start when processing was under way, but the local electricity supply could not cope and expensive diesel had to be used instead. The plan called for the molasses from several sugar factories to be used but the cost of transporting them was too high. Altogether, production costs were far more than the ethanol could have saved and the scheme had to be abandoned after spending £40 million.

QUESTIONS

4.21 The scheme in Kenya failed for reasons which were not biological. Give two non-biological problems mentioned in the text which contributed to the failure.

4.22 What incentive was used in Brazil to encourage people to switch?

4.23 What problems did the introduction of gasohol cause the car owners and manufacturers?

4.24 If cars can run on pure ethanol, why is 3 per cent petrol blended into it?

4.25 Explain why Brazil cannot rely entirely on molasses to supply the ethanol programme.

SUMMARY

Natural waters and soil contain a variety of living organisms including micro-organisms which degrade organic material. The numbers of micro-organisms in natural waters are low but if there is organic contamination of the water the numbers rise and different species are found.

Water purification is designed to remove particulate matter, to make it aesthetically acceptable and to reduce the numbers of harmful organisms. Water can be purified by filtering, by holding in reservoirs or by further treatment and chlorination. Effluent water is treated to reduce BOD, suspended solids and harmful organisms before it is returned to water courses.

Sewage may be treated in several ways, all of which use micro-organisms to degrade organic material in aerobic conditions. Sewage sludge and other wastes can be used to generate fuel. Micro-organisms can be used to remove substances which cause environmental problems.

Chapter 5

INDUSTRIAL MICROBIOLOGY

LEARNING OBJECTIVES

After studying this chapter you should be able to:

1. explain how micro-organisms and cells are grown on a large scale;
2. understand some of the problems of growing organisms this way;
3. explain why cells and micro-organisms can be used to make products;
4. describe the production of a range of products.

5.1 PUTTING CELLS TO WORK

Micro-organisms and isolated cells are used in industrial processes. Using micro-organisms to make food and fuels was explained in earlier chapters; this chapter is about other products made microbially. Industries which use micro-organisms or enzymes, such as brewing, retting flax and tanning, were well established before the start of this century, and within 30 years enzyme and antibiotic production got under way. In the last 30 years, the production of agricultural chemicals, vitamins, vaccines, and flavourings have become major industries. All these products are difficult to make chemically from raw materials, but can be produced by living cells.

The products have huge sales worldwide and the newer biotechnology industries are rapidly adding to the total – biotechnology sales amount to hundreds of millions of dollars each year. One of the earliest genetically engineered biotechnology products on sale was a form of human hormone called Humulin offered by Eli Lilley in 1984. Now several companies offer the hormone; one small vat produces as much hormone as thousands of pituitary glands, and far more cheaply. The 1987 figures from the US Office of Technological Assessments estimated that more than 200 new patents for biotechnological products were issued, with many more waiting in the pipeline. Many of the newer products are medicinal or diagnostic agents, but the list covers a wide range of activities.

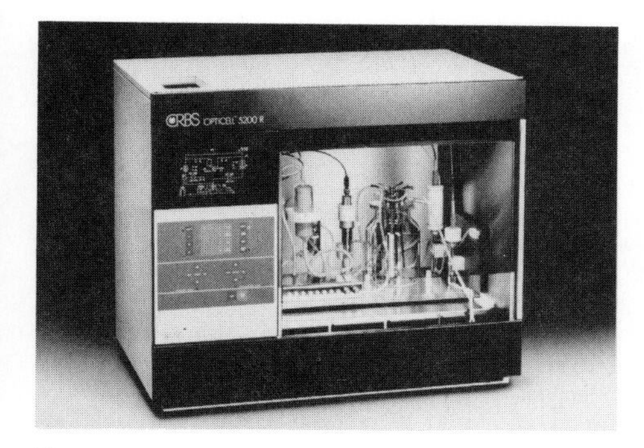

Fig 5.1 Laboratory fermenters are small, holding about one litre of microbial culture. Temperature, pH, and oxygenation are controlled in experiments to investigate the optimum growth requirements for the synthesis of a particular product.

Micro-organisms are grown as animal feed supplements, as sources of enzymes and a variety of chemicals, or to carry out chemical conversions. Some products are part of a cell's normal metabolism, but the ability to make other products may be engineered into cells. For example, vaccine manufacturers use genetically engineered cells to produce the chemicals in harmful bacteria which provoke the immune response. The advantage of using micro-organisms is their very rapid metabolic rate, quickly using the substrate available and turning it into product. There is a wide range of reactions available, and the adaptability of micro-organisms enables them to grow in a variety of environments. Plant cells, too, can be exploited to produce useful materials, but work in this area is only just beginning.

Many processes are developed in laboratories, but not all go into industrial production because there are many problems associated with going up in scale from small quantities in a laboratory fermenter like that in Fig 5.1 to growing the same organisms in the hundred thousand litre tanks seen in Fig 5.2. It is easy to sterilise flasks and a few litres of medium; it is much harder to keep the working part of a factory, the **plant**, sterile.

If we want to use an organism on a large scale we need to know a lot about its growth and needs and how we can influence its metabolism to make the desired product. The product must be exactly the same as when it is made on a small scale, especially if the product is to be used for people. It must not have harmful effects on the people who use it or come into contact with it. Any processing needed to make a consistent final product adds to the expense of the process.

5.2 GROWING CELLS ON A LARGE SCALE

Cells are grown in fermenters which are large tanks holding up to two hundred thousand litres. A simplified diagram of a fermenter is shown in Fig 5.3. The cells are supplied with nutrients and the environmental conditions are carefully controlled to keep them in the desired growth stage. Many processes produce acids which corrode ordinary steel so the tanks are usually made of stainless steel, but occasionally special alloys have to be used if particular metal ions are produced which interfere with the product. Nutrients and other materials are fed into the fermenter by valve-operated pipelines; the rate of entry is monitored and controlled to regulate cell growth. The interior of the tank and all the pipework must be sterile before production begins. The usual way of washing out and sterilising is to flush through with steam which will kill organisms and most spores. Sometimes parts of the system may get contaminated during

Fig 5.2 Industrial fermenters hold thousands of litres. All the ingredients needed in the fermenter are fed in through pipes, sterilised with steam and regulated by valves. There are sample lines to allow the product to be tested and outflow pipes to empty the vessel when fermentation is completed.

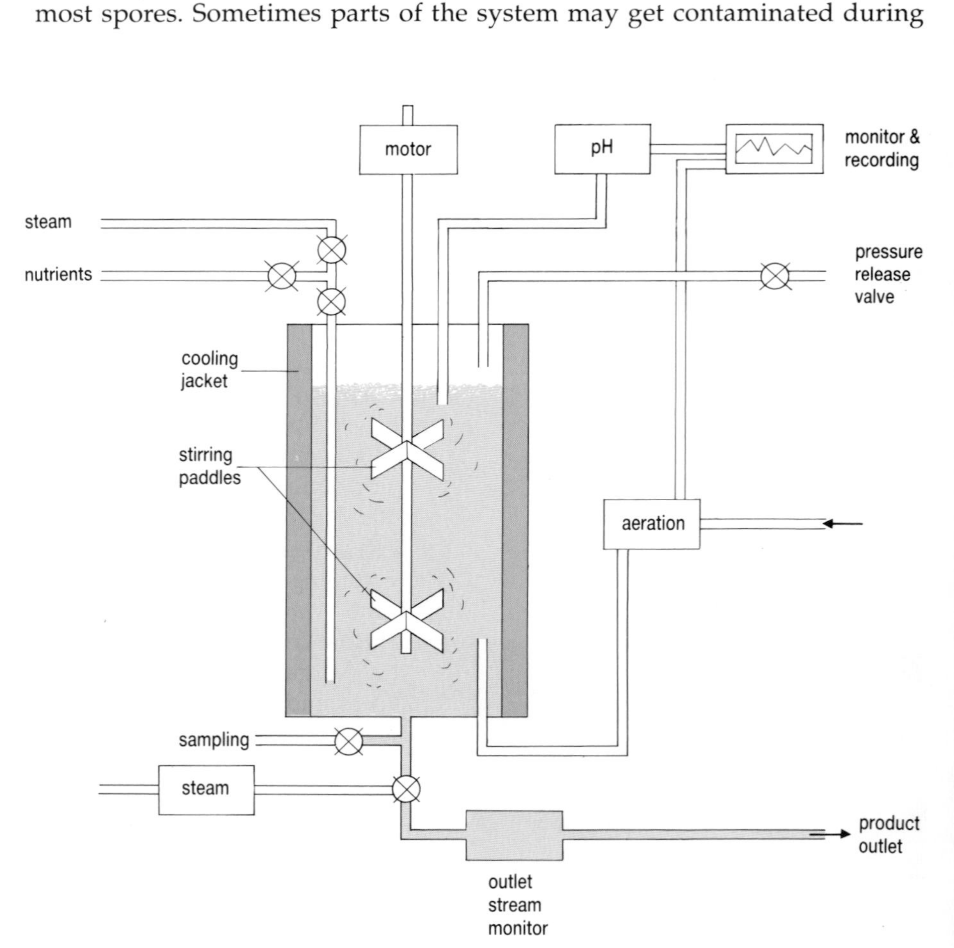

Fig 5.3 A simplified diagram of an industrial fermenter.

processing so the pipework and valves are arranged to enable parts of the system to be isolated and sterilised without having to shut down the whole fermenter.

Industrial fermenters

The interior of the fermenter is monitored by sterilisable **probes** which record temperature, pressure, stirrer speed, pH, oxygen and carbon dioxide levels. These are recorded and electronic control systems with automatic valves regulate them. For example, if the medium becomes too acidic for optimum growth, bases can be added to the tank from a reservoir to correct the pH.

Though the production of materials by microbial activity is called fermentation it isn't often fermentation as defined in Chapter 1. The term is used to mean almost any process using a micro-organism to make a product, including aerobic reactions. Many important organisms are aerobic, so the fermenter has to be well aerated to get maximum growth. Incoming air is filtered, then pumped into the base of the fermenter. It bubbles up through the mixture, and a valve releases the pressure at the

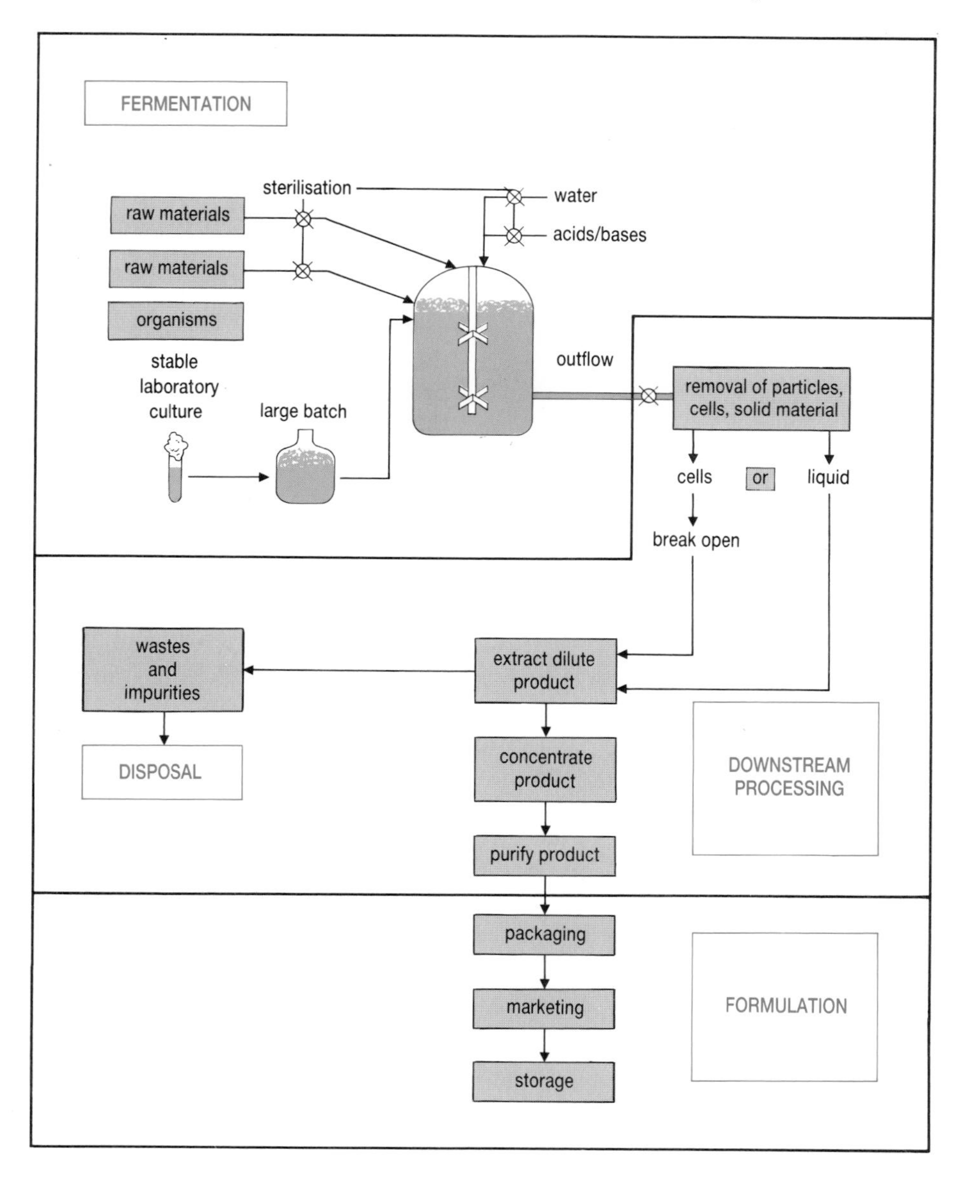

Fig 5.4 Flow chart of a biotechnological process.

top of the tank. Outgoing air is filtered too, particularly if the organisms are potentially harmful. The disturbance caused by the movement of air bubbles through the tank is often enough to circulate the contents to ensure an even distribution of nutrients, but some mixtures, particularly thick or sticky mixtures, need mechanical stirring by a motor driven paddle, called the **impeller**. Anaerobic fermentations are carried out in the same way, but the oxygen concentrations may have to be reduced if the organism is sensitive to oxygen.

The mixture of organisms and nutrients may need preliminary warming before fermentation starts. This can be done by blowing steam at a controlled rate through the contents of the tank. However once the process is started a **cooling system** is vital. An antibiotic producing fermentation may use a tonne of sugar every two days. Though the organisms use much of the available energy, there is still enough left over to raise the temperature by 1°C an hour or more; more heat comes from the paddles and other mechanical movements. The activity in the fermenter can quickly generate temperatures of 60°C or higher which can kill vegetative cells and denature enzymes, bringing the process to a halt. Protein products may also be affected by the heat, so temperatures have to be kept down. Cooling can be by a water jacket round the fermenter or cooling coils inside.

Special precautions are taken if the fermenter holds potentially dangerous organisms such as disease-causing organisms grown to make vaccine, or genetically engineered organisms which may be subject to special containment rules. All entry and exit points, observation and sampling ports have to be sealed to prevent their escape.

The process

A flow chart of the manufacturing process can be seen in Fig 5.4. Each product will need slightly different treatments but the general process is the same. There are three main areas: the fermentation, the downstream processing needed to extract a pure product, and packaging the product so that it has as long a shelf life as possible. Each of these areas has its own problems which must be solved before attempting large-scale manufacturing.

The fermentation

Most processes are batch cultures. A sample is inoculated into a suitable medium in the laboratory, then a larger batch grown on. The organisms are from robust and genetically stable cultures which will not change during the fermentation or between batches. Stocks are usually kept freeze-dried until needed.

At the start of the process, the nutrients and other raw materials are sterilised then fed into the fermenter. The fermenter is then brought to operating conditions and the inoculum added. When stirred, some media make stable foams because of surface-active chemicals in the broth. The foam may block the air vent and any cells caught in it have few nutrients. To solve this, many media have **anti-foaming agents** added to them.

Once the cells or organisms have reached the desired growth phase conditions are kept as close to the optimum for product production as possible. Bacteria grow as single cells dispersed through the medium, but fungal hyphae spread through the medium making it viscous and difficult to stir. However, if they are stirred vigorously enough, mycelia break up into pellets, each pellet acting as a single manufacturing unit.

Eventually the amount of product made declines and the fermenter is shut down, cleaned and sterilised. If the product is an extracellular compound the culture medium can be drained off during growth and fed into the next stages; if it is an intracellular product it has to be harvested when batch growth stops.

Some products are made using a continuous culture system where there is a steady nutrient input, and spent medium, containing product, is drained off. In fact not many products are made this way, though industry would prefer it if they were. Continuous culturing enables manufacturers to run their plant continuously, which is more economical than having to stop and start with the plant idle for two or three days a week. It also ensures that there is a continuous supply of product.

Some products are made in multi-stage processes involving activities in different fermenters. In these the fermenters can be linked together or the processes organised so that batches all finish at the same time to keep machinery in use as much as possible.

Downstream processing

When fermentation is over the product is extracted and concentrated. Whichever methods of extraction and concentration are used they will add to the cost and may affect the nature or stability of the product. Cells are first separated from the medium, usually by centrifuging. If the product is dissolved in the medium it will be dilute and will next need concentrating. The method of concentration depends on the product, for example heating to drive off water cannot be used if the product is a protein because heat denatures proteins. Osmosis, pressure, adsorbant columns or electrical fields are all used. Some very valuable products, for example interferon, which form less than 1 per cent of the crude mixture, can be extracted using monoclonal antibodies in immunosorbant columns which purify and concentrate at the same time. Immunosorbant columns are described in section 5.16.

QUESTIONS

5.1 Why are products such as enzymes and vitamins produced microbially?

5.2 List four environmental factors to be controlled in a fermenter.

5.3 What does an impeller do? Is there any alternative to an impeller?

5.4 Briefly outline the sequence of events between growing the freeze-dried organisms and a bottle of product leaving the factory gate.

5.5 Though most processes are batch cultures, most manufacturers don't like them. Give two reasons why they are unpopular.

5.6 Describe two things that a manufacturer can do to prevent contamination of a product.

5.3 THE POTENTIAL OF CELLS AS SOURCES OF PRODUCTS

The advantage of using living organisms to carry out chemical reactions is that they do them quickly at environmental temperatures. The same reactions in a laboratory may need high temperatures or pressures. For example, micro-organisms convert nitrogen to ammonia then nitrate at 15°C, whereas the Haber process to make ammonia needs temperatures of about 450°C, high pressures and a catalyst.

The DNA in a cell governs the synthesis of the **enzymes** that catalyse cell reactions. Some enzymes in a cell are made all the time and are used continually; other enzymes are made, or **induced**, when they are needed. Enzymes are usually active inside cells but some are secreted through the cell surface to act outside it; for example some amylases are secreted to degrade food materials. If the cells are provided with the right substrate these enzymes may accumulate outside the cells and can be harvested. Some intracellular enzymes catalyse the synthesis of cell growth materials such as polysaccharides, and cells have to be broken open to extract the enzymes or products. Other intracellular enzymes control the production

Enzymes produced by cells, primary and secondary metabolites can be harvested from cells and micro-organisms as commercial products.

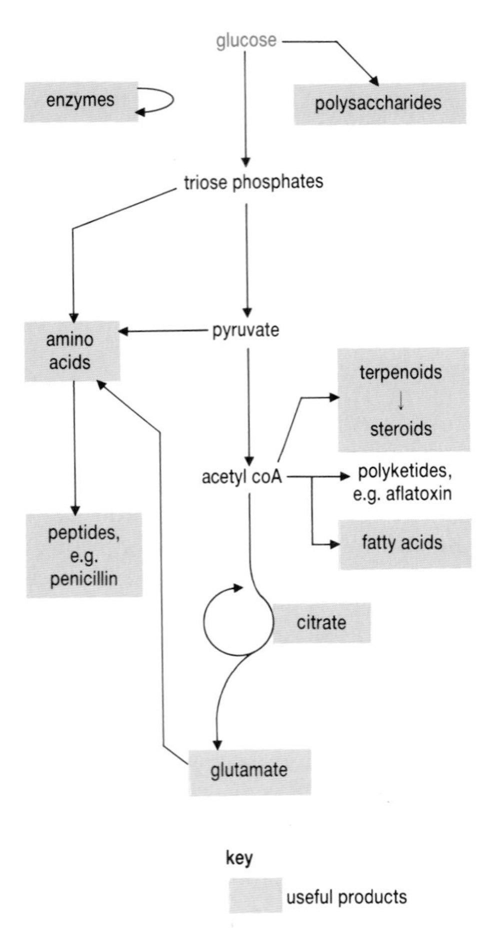

Fig 5.5 Some products made by fungi from glucose and its breakdown.

of chemicals which are active in biochemical pathways, or are precursors for reactions in cells, for example vitamins and citric acid. Enzymes are useful industrially as they are very specific in the reactions they catalyse, with few by-products. A production process using enzymes is cheaper because it runs at low temperatures and purification of the product is much cheaper.

Precursors and the chemicals needed in reactions are called **primary** or **intermediate metabolites**. They are generally used up as quickly as they are made, so special techniques have to be used to obtain large quantities. Some cell products are made only at particular stages in the life of a culture, for example penicillin is only made when the nitrogen content of the substrate falls and *Penicillium* growth has slowed. These compounds, which are not vital but have a useful role, are called **secondary metabolites** and may accumulate in large quantities. Useful secondary metabolites include quinine, codeine, capsaicin (the hot ingredient in chilli pepper) and pigments such as indigo. Fig 5.5 summarises some products made by fungi from the main respiratory pathway and shows the wide potential for exploitation of this pathway.

Compounds such as citric acid are very difficult to make from inorganic chemicals but are made by living cells quickly and easily. Some substances, particularly pharmaceuticals, can't be synthesised at all; these have to be extracted from living cells.

Once the regulatory mechanisms controlling cell activity are understood they may be manipulated to make cells produce large quantities of the compounds we want. Primary metabolites are a problem as they do not normally accumulate. The only way to change this is to upset the cell's normal control systems so that the metabolite is made, but not used. This requires very detailed information about cell activity, and manipulation has to be tightly controlled so that other vital activities are not affected. Occasionally the DNA has to be worked upon to get repressed genes expressed, to induce enzymes, or even to engineer new genes into cells. Fig 5.6 outlines one way of transferring genes from one cell to another so that the recipient cell can produce a compound it could not previously make.

QUESTIONS

5.7 Give one reason why cells may secrete enzymes into their surroundings.

5.8 Explain the difference between a primary metabolite and a secondary metabolite.

5.9 Read Section 5.3 through carefully and give one example of an extracellular enzyme, a primary metabolite and a secondary metabolite that are microbial products.

5.4 WHOLE MICRO-ORGANISM PRODUCTS

Bacteria and fungi can be grown on many substrates to make products. Food manufacturing processes and waste management, covered in Chapters 3 and 4, use whole cells to carry out chemical reactions and bring about changes. Other processes use microbial cells grown on different substrates to make products too, for example *Nocardia bacteria* and *Rhizobium nigricans* are used to chemically alter steroids. This kind of process, done in a fermenter, is called a **biotransformation**. The products include the hormones oestradiol and testosterone, and hydrocortisone derivatives. These steroids are found in animals in tiny quantities so it is impractical to use them as a source of steroids for drugs. Biotransformations of plant sterols, which are common and cheap, are the only economical ways to make quantities of these drugs. A single enzyme, often found in only one or two species, carries out the transformation. For

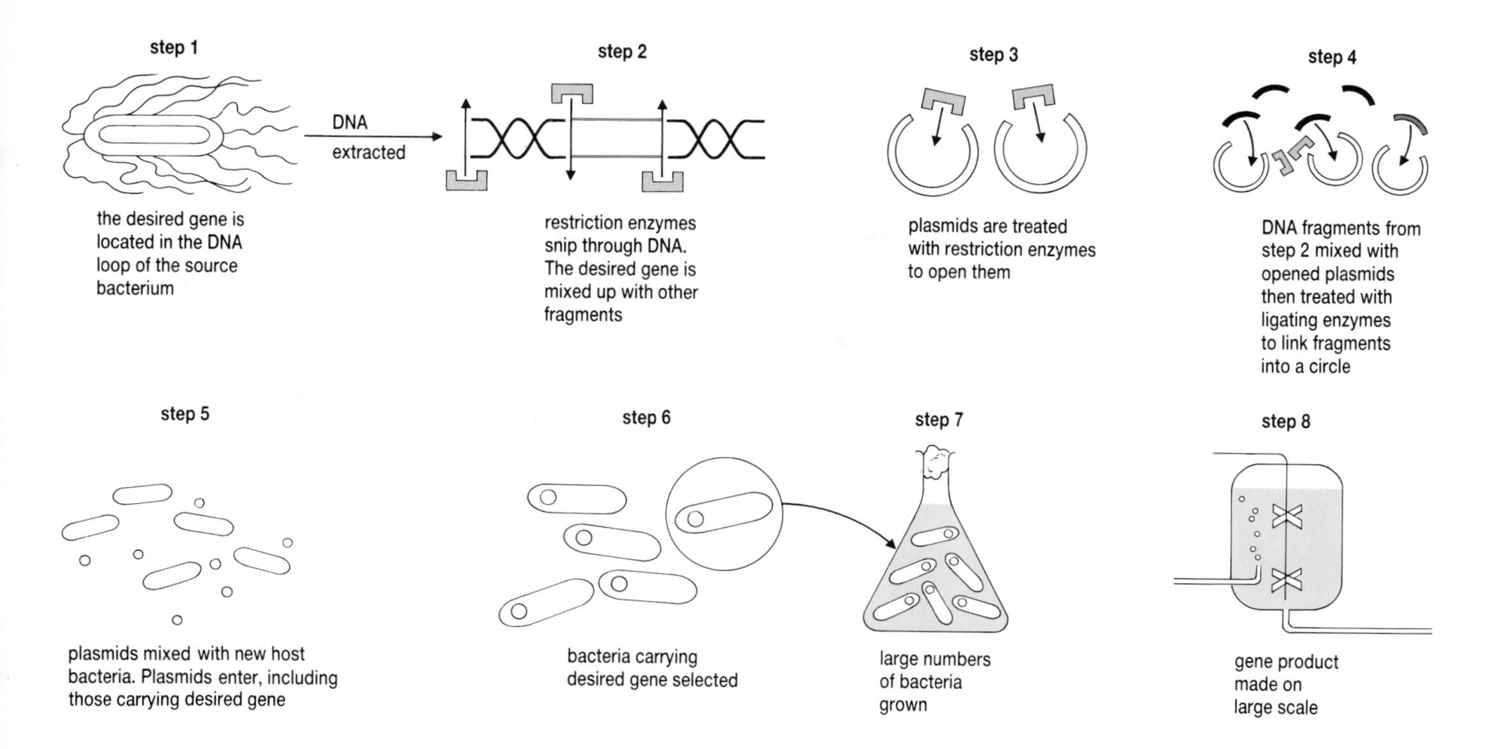

If human genes are used they are not usually snipped out of a cell. mRNA to the desired product is isolated and reverse transcriptase used to make a complementary copy of the gene which is then linked to plasmids

Fig 5.6 Transferring genes.

example progesterone is acted on by an *Aspergillus* species to make an intermediate that can then be chemically converted to cortisone.

The cells used in a process can be immobilised in an inert matrix; yeast in particular is amenable to this treatment. The substrate is then trickled over them and their enzymes act on the substrate to produce the desired compound. This is extracted from the medium as it drains away from the immobilised cells. The technique gives a product uncontaminated by cells, and the cells have a longer useful life.

Occasionally, the bacterial cells themselves are the product. Some bacteria and viruses could be as **biological controls** of insect pests of plants. *Bacillus thuringensis* infects the caterpillars of butterflies and other members of the Lepidoptera. The bacteria produce a toxic protein which affects cells lining the gut making it leaky. As a result the blood of the caterpillar becomes alkaline and it dies. There are several varieties of bacteria which parasitise different insect species. *B. thuringensis* is being researched as a pest control agent, either as an inoculum for greenhouses or as a solution of toxin. Some other micro-organisms, particularly viruses, are being developed to control insect pests. In many crops productivity is limited by available nitrate supplies. Crops such as clover enrich the soil nitrite supplies because of the activity of **nitrogen-fixing bacteria** living in nodules on their roots. The bacteria fix nitrogen from the atmosphere in a form that can be used by plants, enabling them to grow more vigorously. The bacteria most closely associated with plants are *Rhizobium* and its relatives. Each species of bacterium will only grow in its particular host plant. If the bacteria are not present in the soil the host plant grows as well as any other, but if the right bacteria are added it will flourish. Farmers and horticulturists can now buy cultures of *Rhizobia* shown in Fig 5.7 to mix with seeds before planting to improve crop yields.

Fig 5.7 Farmers can buy *Rhizobium* cultures to use when growing leguminous plants. These bacteria fix atmospheric nitrogen and later make it available for their host plants.

Two processes have been chosen to show the use of whole microbial cells; in these processes the cells themselves are the ultimate product.

5.5 VACCINE PRODUCTION

Pathogenic micro-organisms carry chemical groupings, or antigens, which stimulate host immune systems during infection. Vaccines work by presenting a suspension of killed microbial cells, or degraded cells, a weakened strain or a toxoid to an individual to provoke an immune response and protect against further infection. More about the immune response can be found in Chapter 6. Large numbers of pathogenic organisms are grown to make vaccines. There is always a minute risk that organisms may survive the process, or if a weakened strain is used that it may suddenly regain its ability to cause disease, and the risk is larger to the people who work in the plant making the vaccines. The problem is reduced, however, if genetic engineering techniques can be used.

Genetic engineering is used increasingly in vaccine production. The genes responsible for antigen production in some disease causing organisms are isolated and transferred into less dangerous organisms such as *E. coli*. The engineered organism is cultured and makes the antigens encoded in its new genes which can be extracted free from cell debris. There is much less risk to the workforce and the recipients of the vaccine should suffer less side effects from the presence of other cellular materials. The main difficulty has been the identification of the antigens most important in the immune response and the genes which encode them. Vaccines of this sort are in use to prevent loss of young farm animals from diarrhoea, and are being trialled against malaria using an engineered *Salmonella* strain carrying the surface protein from *Plasmodium falciparum*. Similarly, engineered yeast is being investigated for use against hepatitis B and new acellular vaccines are being produced against whooping cough, typhoid, and sheep foot rot.

Manufacture

The organisms for a bacterial vaccine are grown in a fermenter and provided with the optimum conditions for antigen and toxin production. Viruses are grown in sheets of animal cells in serum-free tissue culture. As most of the organisms are dangerous the precautions to contain them are particularly rigorous, for example the culture may be stirred by a magnet controlled from outside; an ordinary stirrer shaft would have to pass through an exit that could be a source of leaks. Workers are well protected, as can be seen in Fig 5.8.

Fig 5.8 Process workers growing diphtheria bacteria to make vaccine are well protected.

After the organisms have grown they are harvested in sterile conditions. Bacterial cells are separated from the culture medium by centrifuging. Further processing depends on what is used to make the vaccine. The cells may be used as whole cells, live or dead, or component antigens may be extracted by breaking open the cells. Antigens can come from the cell wall of the organisms, for example cholera and whooping cough, or from the capsule. Some organisms, such as meningococci, produce an antigenic capsule which spreads into the culture medium and has to be extracted from it; others secrete toxins which must be extracted in the same way.

Processing

Toxins, such as those made by tetanus and diphtheria bacteria, are usually treated with formaldehyde to make them harmless but they are still able to provoke an immune response. They have to be purified to remove other microbial materials before they can be used as a vaccine. Extracted antigens are purified using immunosorbant columns.

When vaccines contain several antigens, or antigens from different strains, they are blended to form a standard mixture. Many antigens are made more effective by combining them with an **adjuvant** that boosts their ability to provoke an immune response, such as aluminium hydroxide. Liquid vaccines may need a stabiliser to give them a longer shelf life. If the vaccine is made from live weakened strains, the cells are freeze-dried to preserve them (see Fig 5.9(a)). The vaccine has to be reconstituted before it is administered. The final vaccine is checked for sterility and effectiveness, and that it is safe to use.

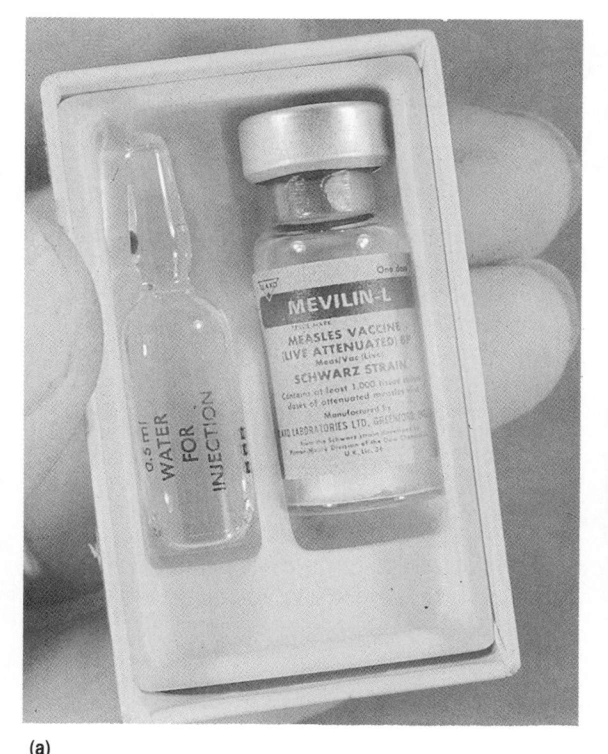

(a)

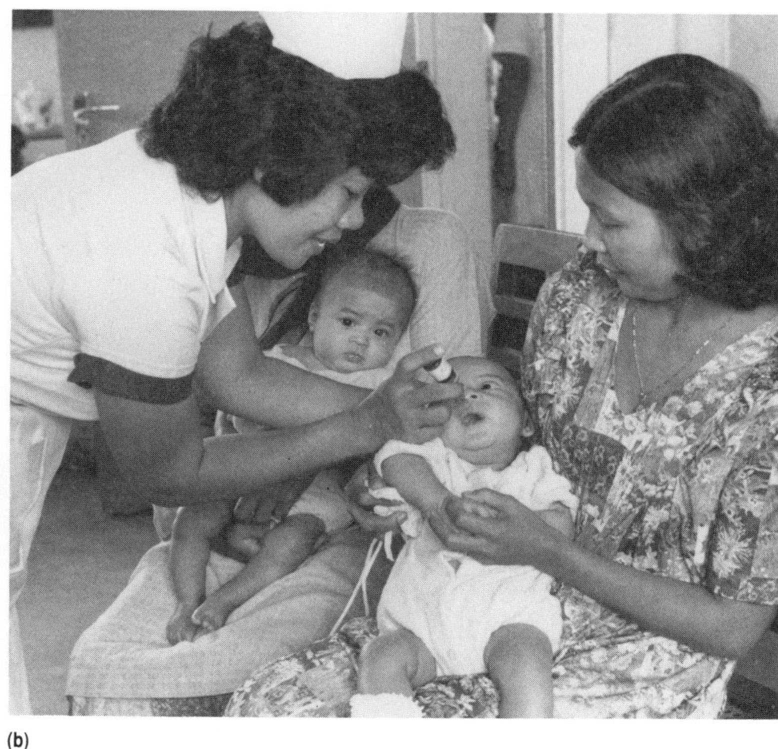

(b)

Fig 5.9 **(a)** Vaccines may be dehydrated to lengthen their storage life; they are diluted with sterile water before injection.
(b) Not all vaccines have to be injected. Some, like this polio DPT vaccine, can be absorbed from the gut, which makes vaccination a pleasant experience.

5.6 SINGLE-CELL PROTEIN

Micro-organisms are used to make protein rich materials for people and as animal feed. The processes frequently use wastes from other industries as a nutrient source for the micro-organisms. The use of fungi to make a human protein food was explained in Chapter 3, but there is another important process which use bacteria in a specially designed fermenter. Bacteria are used to make animal feed SCP from methanol which is a by-product of North Sea gas production; the product is sold as **Pruteen**™.

To make Pruteen, the organism *Methylophilus methylotrophus* is grown aerobically on methanol dissolved in water with mineral salts and ammonia. A special fermenter, known as a pressure cycle fermenter, was developed to grow the organism. It runs continuously and uses sterile compressed air to circulate the contents, (see Fig 5.10). This reduces potential contamination as there is no paddle shaft entry point. The fermenter is sterilised by steam, then sterile air and sterile water are passed through. Sterilised nutrients are fed into the fermenter together with a few litres of microbial culture. After 30 hours the number of cells reaches the working density and they are continuously drawn off while the medium is maintained. The cells are flash dried and the liquid recycled. The cells are broken up to improve the digestibility of the product, and the pH and mineral balance are adjusted before final formulation and packaging.

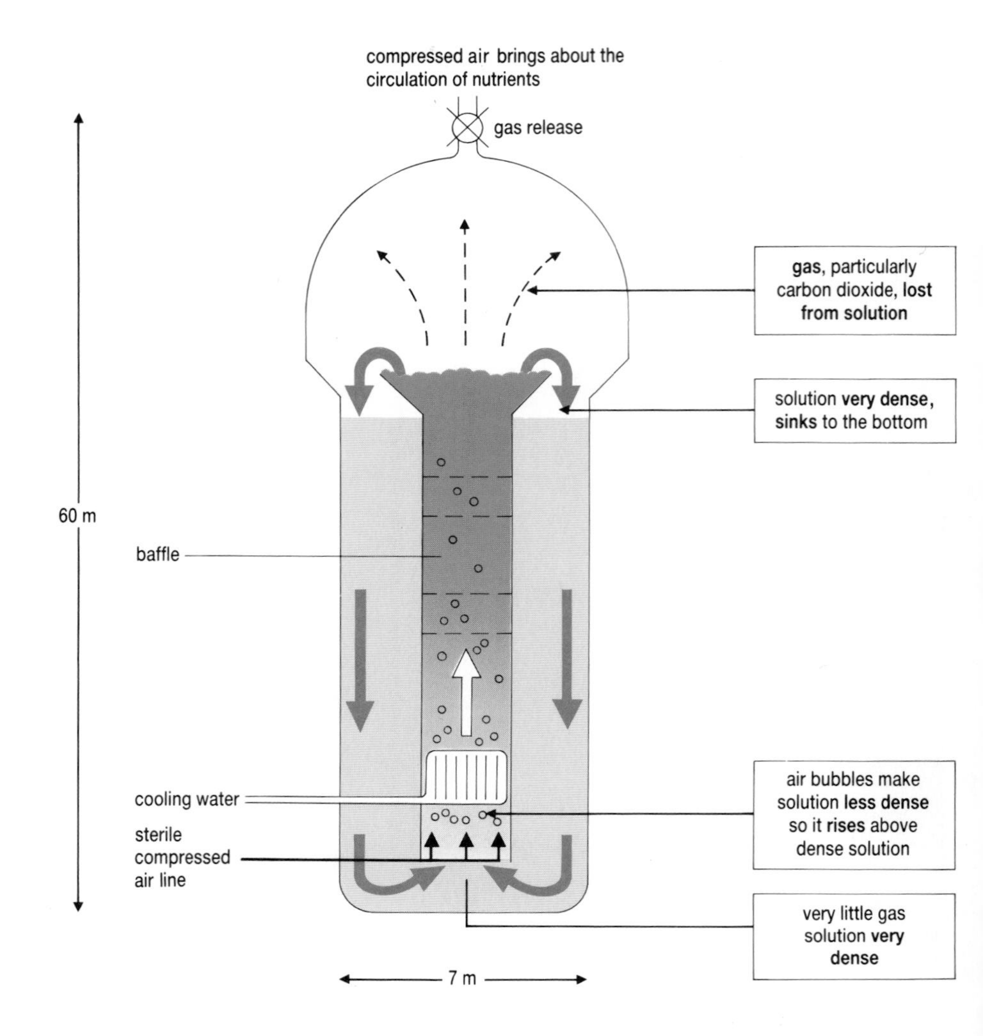

Fig 5.10 A pressure cycle fermenter.

QUESTIONS

5.10 Why are microbial antigens needed in vaccines?

5.11 Give two advantages of engineered vaccines over ordinary vaccines.

5.12 Draw a flow chart outlining the sequence of manufacture of a vaccine made from antigens extracted from whole cells of several strains of a pathogenic organism.

5.13 Construct a similar flow chart for the manufacture of Pruteen™.

5.14 Bovine somatotrophin (BST), a growth hormone made microbially, was used in anonymous trials in Great Britain to monitor the effects on milk yield. Many people were very upset at not knowing whether their milk had come from cows treated with BST. Why do you think people were worried? Would it have made any difference to the trials if people did know which herds were being treated?

5.7 CHEMICALS MADE BY MICRO-ORGANISMS

Microbial cells release some useful chemicals into their environment, for example enzymes and antibiotics. Other useful compounds accumulate inside cells and can be extracted by breaking the cells open. Selection and genetic engineering has produced strains of organisms which synthesise far more of these substances than normal. Three microbial products – enzymes, antibiotics and citric acid – have been chosen to illustrate the ways micro-organisms can be used to make products, and the areas that use microbial products. There are other equally important processes, and these are also mentioned briefly.

Microbial vitamins, which have sales second only to antibiotics, and amino acids are more easily made microbially than by chemical methods. There are two amino acids that are made in large quantities. Lysine and glutamate (as MSG) are used to enrich cereal-based animal feed and to improve the flavour of processed foods.. The bacterial strains used to produce these amino acids make more than they can use, and altered cellular permeability allows the excess to be released into the surrounding medium.

Table 5.1 Products made by micro-organisms

Substance	Organism used	Use
alginate	*Azotobacter, Pseudomonas, Xanthomonas*	thickener and stabiliser in ice-cream; instant puddings; paints; textile dressing; spray for Christmas trees; immobilising cells
avermectins	*Streptomyces avermitilis*	worming farm animals; removes ticks from coat
citric acid	*Aspergillus niger* on molasses syrup	flavouring; drinks; jam
curdian	*Alcaligenes*	gelling agent in low-cal soups and puddings
dextran (1,6 glucan)	*Leuconostoc mesenteroides, Klebsiella*, etc. on sucrose	medicinal use; blood plasma biochamical adsorbant in Sephadex™
insulin	*Escherichia coli*	medical use
lactic acid	*Lactobacillus helveticus* on malt	flavouring; confectionery; soft drinks
lysine	*Brevibacterium flavum*	enriching bread
monosodium glutamate	yeast on sugary substrate, *Corynebacterium glutamicum*	flavouring
nucleotides	yeast on sugary substrate	flavouring; soup; sauces
propylene oxide	*methylococcus capsulatus*	making alkene oxides to polymerise into polypropylene and other plastics
pullulan	*Pullularia, Aureobasidium*	coating covering foods
vitamin B_2	yeast	enrichment of food
vitamin B_{12}	*Streptomyces* on malt	dietary supplement
vitamin C	*Gluconobacter* convert sorbitol to sorbose as a step in the chemial manufacture. *Pseudomonas* also used	enrichment of food; sorbose used as a suspending agent for drugs
xanthan gum	*Xanthomonas campestris*	thickener in sauces; cheese; cosmetics; paint; ink; instant puddings; thickener of water for oil extraction; stabilises emulsions

A sweetener has been recently introduced, **aspartamate**, which is sold under brand names such as Nutrasweet™. It is made from the amino acids phenylalanine and aspartic acid. These can be made microbially, but current yields need to be improved.

A plant hormone is made in quantity by a fungus, *Gibberella fujikuroi*, which is a plant pathogen. It releases gibberellic acid, a hormone which causes stem elongation, one of the symptoms of infection. The fungus is now grown in batch culture to produce gibberellic acid, which is a secondary metabolite produced when fixed nitrogen levels fall in the medium.

Microbial metabolites can be extracted and packaged as useful products.

There is a lot of interest in human cell products as there is a large market for medical use. Animal cells are difficult to grow on a large scale in a fermenter so researchers are trying to engineer human genes into bacteria for industrial production. Human insulin was one of the earliest of these products. It is made by *Escherichia coli* carrying the gene for pro-insulin, an inactive form. The bacterium secretes the hormone into the culture medium from which it is extracted and treated to make it active. Insulin is now made in large quantities as are other human hormones.

Table 5.1 summarises some of the other microbial products.

5.8 ENZYMES

Fig 5.11 Beads of immobilised enzyme which can be immersed in a substrate solution to catalyse a reaction then can be removed and reused.

Microbial enzymes are used to catalyse a range of reactions at low temperatures and pressures. Enzymes have been used commercially for the whole of this century, originally in the textile and tanning industries, and, surprisingly, the first biological washing powder was formulated in 1913. However, the large-scale production of enzymes from micro-organisms is a more recent development. Enzymes are used by industries as diverse as textile manufacturing, food processing, diagnostic kits and materials production. The efficiency of the enzymes in use is improved by good design and engineering of the system.

Properties of enzymes

Industrial enzymes must have a long shelf life and stand up to rough working conditions. The organisms grown are selected for their robust enzymes that can work in a wide range of conditions. Most work in a temperature range of 8–55°C with optima between 30 and 50°C. They may have a wide pH tolerance or may work in the presence of chemicals that usually inhibit enzyme action such as sulphur dioxide or polyphenols. Some enzymes have exceptional properties, for example there is a thermostable α-amylase that can degrade starches at temperatures over 100°C. They are widely used to convert cheap carbohydrates into fructose and glucose, in washing powders and low phosphate detergent washing up liquids, and to replace animal rennet in the dairy industry.

Enzymes are more efficient and have a longer working life if they are immobilised on inert supports. The product is free of potentially troublesome contaminants.

Enzymes are used as solutions added to the substrate, but this has disadvantages as the enzymes get washed out of the reactor and are diluted by the material they work on. It is better to use them attached to inert materials, or **immobilised**, on ceramic or polymer gels or on membranes, or encapsulated. They are more stable used this way, which is similar to the way they act in cells where most enzymes are firmly attached to membranes and only part of the molecule exposed, and their life is prolonged. A batch of immobilised enzyme can be seen in Fig 5.11. Immobilised enzymes are dipped into the substrate which is circulated to give maximum contact with the enzyme, and the product is free from contaminating enzyme molecules. Immobilised enzymes are ideal for use in continuous processes.

Table 5.2 Some enzymes produced by micro-organisms

Enzyme	Use	Organism
α-amylases	breakdown starch to dextrins in mash; some produce maltose; degrade stains on clothes; paper manufacture; desizing starches on textiles; thicken canned sauces; improve flour by degrading starch to sugar; reduce the rate of bread going stale; help separation of loaves; vegetable juice extraction; formation of glucose syrups from carbohydrate	*Aspergillus oryzae, Bacillus subtilis, B. licheniformis, Actinoplanes*
β-glucanase	degrades β-glucan in beer; prevents pipe blockage and cloudiness	*A. oryzae, B. subtilis*
catalase	preservative in soft drinks; oxygen for foam rubber making	*A. niger*
cellulases	washing powder colour brightener; animal feed from rye grass and straw	*Trichoderma* spp, *Penicillium, Aspergillus*
amylo-glucosidase	breakdown starch and dextrins to glucose	*A. niger, Rhizopus*
glucose isomerase	converts glucose to fructose; soft drink sweetener; fillings and icing; for cakes; jams	*B. coagulans, Streptomyces*
glucose oxidase	preservative in soft drinks; detects glucose in diabetics' blood	*A. niger, Gluconobacter*
lactases	conversion of lactose to galactose and fructose for sweetened milk drinks; conversion of whey to sweetener foods for lactose-intolerent people	*Kluyveromyces lactis*
lipases	accelerate ripening of certain cheeses	*Pseudomonas, Candida, Aspergillus*
pectinase	clarifies wine and juices by degrading pectin; releases juice, oils and colour from fruit and citrus peel extracts for soft drinks	*A. awamori, Erwinia*
proteases	break down protein stains on clothes and food residues on dishes; degumming silk without damage; flour improvers; degrade gluten in strong flour for biscuit making; alteration of milk and whey proteins; bating leather to make it pliable; removal of hair from animal hides for leather; soya bean products; fruit drinks; meat extracts and preparations; gelatin; drug assay in blood	*B. subtilis*
pullulanase	manufacture of dextrose and maltose syrups for use in soft confectionary; soft ice-cream	*B. pullulans, Klebsiella*
microbial rennet (and chymsosin)	coagulating casein in cheese-making; baby food manufacture	*K. lactis, Mucor miehei, M. pusillus, M. rouxii*
sucrase (invertase)	soft centres for sweets	*A. oryzae, Saccharomyces*
streptokinase	treatment of blood clots and bruises	*Streptomyces* spp

Production

Enzymes are produced by fungi or bacteria in well-aerated batch cultures, except for the production of glucose isomerase by *Bacillus coagulans* which is a continuous culture. The nutrients are similar for different organisms, usually a sterile starchy medium such as barley and soya bean meal or potato starch in a buffered solution. Glucose is omitted because it can repress the production of some enzymes. Many enzymes are secreted into the medium but others remain inside the cells where they normally act.

When growth has finished, after two to five days, the temperature is raised enough to kill the cells but not to damage the desired enzyme, which may be more heat stable than its producing cell. Intracellular enzymes are released by grinding the cells but often whole cells can be used. Particulate matter is removed and the solution is concentrated by evaporation or ultrafiltration. The enzymes may be completely separated from cells but often an impure mixture, which is cheaper to make, is good enough. The enzymes are then formulated and packaged for distribution. Some important microbial enzymes are listed in Table 5.2.

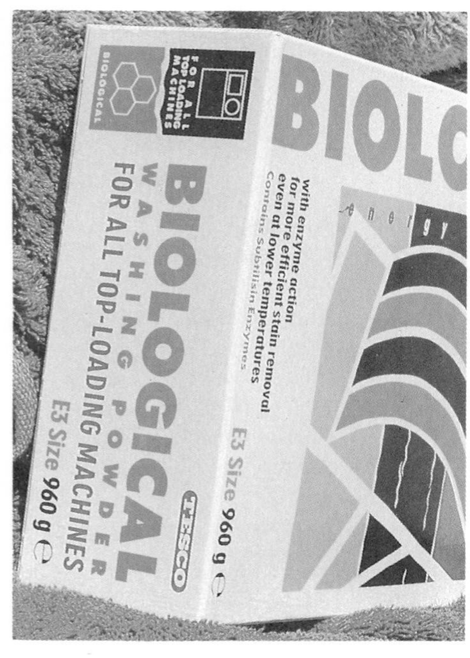

Fig 5.12 'Subtilisin' enzymes are used in this washing powder to loosen and degrade organic material staining laundry. Biological washing powders work best at the lower temperatures needed to

Washing powder enzymes

The wide-scale marketing of biological washing powder began in 1965, though these early powders had to be reformulated as some users and workers developed allergies to the enzymes. Biological washing powders and liquids contain a range of enzymes that can degrade proteins and carbohydrates attached to textiles; they are less successful at removing fats. The enzymes have to be able to work in an alkaline solution produced by detergents and at temperatures between 10 and 90°C.

Concern about environmental pollution and the use of fabrics which must be washed at low temperatures have made washing powder enzymes even more important (see Fig 5.12), since they are effective at low temperatures and work on stains that low phosphate detergents do not cope with very well. A new additive is a cellulase that degrades the microfibrils that form on cotton fibres during washing and wearing. This brightens the colour of the cloth and makes it feel smooth.

After extraction, the enzymes are encapsulated with salts and other materials and coated with an inert substance for safety during washing powder manufacture. The encapsulation protects the enzyme while the powder is made and also from the action of the detergent during storage.

Sugar-releasing enzymes

Many foods have sugar added as a flavour or sweetener. Sucrose, or cane sugar, became expensive so manufacturers looked for cheaper sweeteners. Microbial enzymes are now used to make the sweetener fructose from cheap carbohydrate. Starch can be broken down to glucose by acid hydrolysis, but glucose is less sweet than sucrose and the product may contain undesirable by-products. Fructose, however, is sweeter than sucrose so less needs to be added for the same sweetness, but it is more expensive – unless it is made using microbial enzymes.

A cheap carbohydrate such as corn starch is heat-treated to gelatinise the starch in it and then treated with microbial α-amylase and amyloglucosidase which work at high temperatures. The enzymes degrade the starch to dextrins then glucose. The product can be left as glucose syrup for use in a range of foods and drinks as shown in Fig 5.13, or it can be treated to make fructose. The dissolved glucose is treated with another microbial enzyme, **glucose isomerase**, which converts glucose to fructose. Glucose isomerase is used as an immobilised preparation and has a long life. The product contains a high proportion of fructose so it is as sweet as sucrose and is usually left as a concentrated syrup. The product is so successful that millions of tonnes of starch are now converted to sugar syrups each year.

Fig 5.13 Glucose syrup is an important product used in a variety of foods including this frozen dessert.

Biosensors

Enzymes have a remarkable ability to recognise and select one type of molecule from a mixture of many different types. This ability can be used to detect the presence of certain molecules, even in very low concentrations. In a biosensor, a reaction between the molecule to be detected and an immobilised enzyme molecule causes a change which is transduced into an electronic signal. The amount of chemical in the mixture can be monitored by the number of reactions taking place.

Different sensors work in different ways. The trigger may be the appearance of a product from the reaction, or the movement of electrons during it, or the appearance of light or other factors. Fig 5.14 illustrates the principle on which a biosensor works.

Those on the market and in the pipeline are able to carry out sensitive diagnostic tests that before could only be done in laboratories with specialist equipment. They can also be used to monitor the progress of industrial manufacturing processes. Biosensors have been developed to measure glucose concentration in the blood of people suffering from diabetes and for other monitoring activities such as the detection of hormones, alcohols and pollutants.

(a)

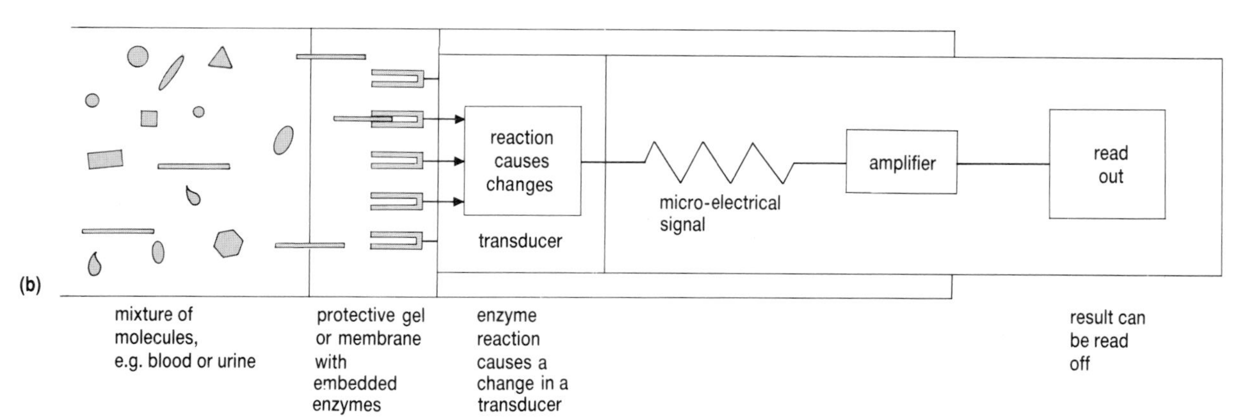

Fig 5.14 **(a)** Diagnostic tests will become much quicker, easier and more accurate as biosensors, like this one for diabetics to monitor glucose levels, become available.
(b) How a biosensor works.

5.9 CITRIC ACID

Some organic acids are traditionally made microbially, for example the production of ethanoic acid, or acetic acid, in vinegar which is dealt with in Chapter 3. Ethanoic acid was at one time made by a species of *Clostridium* as a raw material for plastics and dyes and was the first industrial microbiological product in the 1880s, but other methods of production became much cheaper. Citric acid, properly called tricarboxylic acid, has been made microbially for many years by a much more successful process. Citric acid is used as an ingredient in a wide range of foodstuffs, soft drinks, sweets and processed foods and pharmaceuticals.

Aspergillus niger is grown on waste carbohydrate such as molasses in a fermenter. The fungal cells do not make much citric acid normally; it is a primary metabolite made during the breakdown of glucose to release energy. Even if there is a hold-up in metabolism, the process is inhibited at an early stage and citric acid does not accumulate. Manufacturers therefore had to find a way to make the organism produce excess citric acid by disrupting normal cell control systems, but not to disrupt them so much that the cells could not grow. This could be done in the laboratory by depriving the cells of manganese ions, but in the scaled-up steel fermenters it was impossible to keep manganese out of the medium. The alternative was to grow the organism in medium with excess ammonium ions. These encourage the breakdown of glucose but overcomes the inhibition which stops citric acid accumulating. The excess citric acid diffuses through the cell wall into the medium. When the batch fermentation is finished, the fungal cells are separated out and citric acid extracted from the liquid.

5.10 ANTIBIOTICS

Antibiotics are manufactured as medicines but there are other uses. For example *Actinomyces natalensis* produces an anti-fungal agent sold under the name of Pimaricin. This can be incorporated into the coatings of foods such as cheeses which depend on bacterial action to mature the flavour but may be spoiled by surface mould growth in storage. Pimaricin affects the cell wall structure of fungi but does not affect bacteria or the human consumers.

Once the first antibiotic, penicillin, had been discovered others were quickly identified. Several species of fungi and some bacteria, particularly the Actinomycetes, are now known to be able to synthesise anti-microbial agents, though not all are suitable for use in people or animals. Many antibiotics are uneconomical to manufacture in the small quantities that may be needed, so though there are several thousand antibiotics known, 1500 from the Actinomycetes alone, only about a hundred are available commercially.

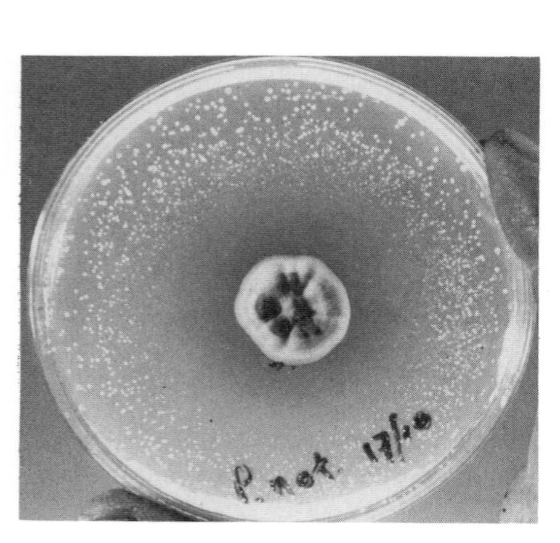

Fig 5.15 Colonies of Gram positive staphylococci bacteria grown on a nutrient agar plate with a culture of *Penicillium chrysogenum* in the centre. Few bacterial colonies have been able to grow near the fungus because it secretes penicillin into its surroundings, inhibiting bacterial growth.

Penicillin

The first antibiotic, penicillin, was originally isolated from a fungus, *Penicillium notatum*. It is still one of the commonest prescribed drugs in the world. It is a secondary metabolite, a type of molecule called a β-lactam, which affects cell wall synthesis of Gram positive bacteria. It is made only when the nitrogen content of the medium falls to a low level. Penicillin was not made in very large quantities by the original strains, often at levels of less than 1 μg for each 2 g of sugar used. *P. chrysogenum*, shown in Fig 5.15, was discovered to have a higher yield and would grow in submerged culture. Several strains have now been selected for high penicillin production.

The conditions in the fermenter are designed to bring the culture to the mature stage where penicillin is produced, and held there. The fungus may convert a large proportion of the medium into penicillin and better extraction techniques have improved the yield. Genetic engineering of the fungus may result in an even greater yield, and make manufacturing easier.

Manufacture

A suspension of conidiospores is inoculated into a starter medium and incubated. The starter is then added to the substrate, which is often waste carbohydrate such as corn steep liquor which contains a precursor for penicillin G, with additives, in the fermenter. The proportions of different sugars in the medium can influence the rate of growth and the amount of antibiotic made. In the fermentation the fungal mycelium is spread as pellets in the nutrient. Oxygen levels are critical in this fermentation as the yield of penicillin is determined by oxygen availability. The medium is maintained at about 24°C and slightly alkaline for optimum growth.

The fermentation takes about a week in all. Penicillin is made and secreted into the medium from about 40 hours onwards when exponential growth has ceased. The fungal mycelium is separated from the medium by filtering and the antibiotic is then extracted from the medium. Solvents are used to purify the antibiotic, which crystallises.

The penicillin secreted by the fungus is really a mixture of compounds; the main one is penicillin G. It can be converted to other forms of β-lactam with slightly different activities. For example penicillin G can be made into ampicillin and other derivatives, and cephalosporins. Cephalosporin is an antibiotic which can be made by micro-organisms but it is cheaper to make it from penicillin.

QUESTIONS

5.15 Refer to the section detailing bacterial cell wall structure in Chapter 1. How does penicillin affect the growth of bacteria?

5.16 Construct a flow chart for ampicillin production.

5.17 Study Table 5.2 then list five different uses of enzymes.

5.18 What is meant by an immobilised enzyme?

5.19 How do the properties of industrial enzymes differ from normal enzymes in living cells?

5.20 Outline a procedure you could use to determine the best temperature to wash out gravy stains with Whizzo biological washing powder.

5.11 PLANT CELL PRODUCTS

Plants are an important source of pharmaceuticals; about a quarter of our medicines and drugs are plant-derived. We also extract flavours, perfumes, dyes, latex and gums, essential oils and waxes such as jojoba, and agricultural chemicals such as pyrethrins. Many drugs, dyes and flavourings are extracted from wild plants rather than cultivated plants but supplies are unreliable because of changing weather conditions, the activity of plant pests, or political conditions in the source country. The plant may even be endangered by over-collection. It is estimated that plant derived pharmaceuticals in the USA alone are worth nearly a billion dollars a year. The top four plant drugs are: steroids for the contraceptive pill from yams, codeine from the poppy, and atropine and hyoscyamine from deadly nightshade for nerve conditions. Perfume manufacturers spend nearly two billion dollars worldwide on essential oils and waxes.

Producing these compounds in culture in controlled conditions would ensure a high quality product that could be supplied all year round irrespective of seasonal or climatic problems. Suspension cultures could be used to produce important compounds, and callus and explant cultures to produce large numbers of important plants. Standard industrial fermenters cannot be used for plant cells for a variety of reasons, though the provision of sucrose in the medium does reduce the need for light. Glass or plastic vessels are used, stacked closely together in racks with subdued artificial lighting. So far few plant cell products are in commercial production because it is difficult to get stable cell lines that will multiply quickly, but many areas are being explored. In contrast the production of plants by tissue culture techniques and micropropagation is a major industry.

5.12 CELLS IN SUSPENSION CULTURES

There are few commercial applications yet for suspension cultures as there are several problems still to overcome. Plant cells are very slow growing and selection for faster growth may not necessarily lead to higher yields. Currently cells in culture tend to produce less than cells in intact plants; the cultured cells synthesise the chemicals in the same form as in the whole plant, but they tend to store the product so it can only be extracted if the cells are broken open. Cells then have a very short useful life which, together with the long generation times involved, make the process uneconomic.

Plant cells are sensitive to rough handling, and are not tolerant of industrial fermenter conditions. The plant cells could be immobilised on inert supports; they prefer to grow close together and the medium can be circulated over them. Investigations show that secreted products can be extracted from the medium as it is recycled, and large numbers of cells can be used. It may be possible to put adsorbtion units in the reactor vessel to compete with the plant cell as a storage site for products.

Plant cells can carry out transformations of chemicals, for example changing steroid molecules. Bacterial cells are used for this at the moment but potentially plant cells could be used also.

Shikonin

The main plant product from suspension culture is shikonin, a pigment for dyeing silk, but it is also used to treat burns and inflammations. The pigment is made by cell cultures of *Lithospermum erythrorhizon*. Selective breeding has improved the yield from less than 1 per cent to 23 per cent of dry weight, but even so it remains expensive at about $4500 per kilogram. Potentially there are many other compounds which could be made this way and processes are being developed or are at the trial or pilot plant stage.

Plant cells do not tolerate the rigorous conditions in microbial fermenters, so other techniques are developed.

The production of the hot ingredient of chilli pepper, capsaicin, from immobilised cells has been well studied but is not yet in industrial production. Cells of the chilli pepper can be grown on inert beads or plastic sponges in columns with nutrient solution trickled through. Chilli is widely used to flavour processed foods, but supplies are irregular and very variable in quality so there is a large potential market for cultured chilli.

5.13 CLONING PLANTS

Producing new varieties

Cell culture techniques are used to produce large numbers of variant plants for crop improvement. Ordinary breeding programmes use huge numbers of plants; in wheat breeding upto 1.5 million seedlings may have been used by the second generation. Far fewer plants need to be fully grown and allowed to set seed using micropropagation techniques. Plant cells which have had their cell walls removed are called **protoplasts** and can be kept in culture like other cells to provide a good source of variation. They will quickly secrete new cell walls and generate new plantlets. Surprisingly, the young plants are often quite different from each other, even though they are all from the same parent. This type of variation is called **somatic variation**.

The technique is very useful for producing variant plants of some species, such as lettuce, which are difficult to pollinate artificially in a cross. It is quick, variants for crop improvement trials take weeks or months to grow rather than years, and fewer favourable gene clusters are lost than in conventional methods. Haploid cells such as pollen can be used to generate lines of plants which are homozygous for every character and are useful parent lines. The details of this process can be found in Chapter 7.

New genes cannot currently be engineered into cells with their walls intact but can be inserted into protoplasts. This could also be a quick route to crop improvement. Hybrid protoplasts from different parents can be made by treating protoplasts with polyethylene glycol. Genetic fusion follows and the hybrid has characters from each parent. If the fusing cells are also treated with UV light a hybrid cell can be obtained which has the nucleus of one cell type and the cytoplasm of another, a **cybrid**. This allows certain cytoplasmic features such as mitochondria to be passed on from one of the parents. It is hoped that the technique can be used to combine cells with different characteristics such as high yield with weather resistance, as well as generating cells from parents which normally cannot interbreed.

Efforts are also being made to isolate desirable plant characteristics and transfer them into cells using plasmids carried in a suitable bacterium such as *Agrobacterium tumifaciens*. New plantlets with the desired characters could then be regenerated from these cells. The technique works well with tomatoes and potatoes but it is difficult to get genes into cereal crops or to regenerate plants from their cells.

Fig 5.16 A tequila farm! In Mexico, the juice of the agave plant is used to make a fiery drink. These cloned plants in a nursery will supplement the slow growing wild agave which is becoming scarcer.

Producing identical plants

Plant cells can be used to generate large numbers of identical plantlets by **micropropagation**. Explants of desirable or rare plants are grown as callus cultures, or as explants, on sterile agar with regulators to stimulate cell division and differentiation. Usually dividing tissues such as buds or shoot tips of a single good plant or seedling are used for the explants. The cells grow and divide for a few weeks and plantlets develop; these can then be used as a source of more tissue for more explants. Very quickly a large number of identical plants, or clones, is built up. The little plantlets that develop are separated and potted on, as shown in Fig 5.16.

Commercial breeders grow a range of horticultural and agricultural plants, quickly producing large numbers of plants with the desired characteristics. Important plants such as yams, the solanums and oil palms are all routinely propagated by micropropagation. Plants which are difficult to obtain as seed, such as bananas, or which are difficult or slow to propagate can now be produced in large numbers. One genus under investigation is *Podophyllum*, which is the source of a drug used in cancer treatment. Many expensive cut flowers are produced this way too, for example gerberas, lilies and orchids were seldom seen in florists but are now extremely common as a result of micropropagation.

ETOPOSIDE

Etoposide is a drug which is obtained from a crude extract of the rhizomes of a genus of plants, the podophyllums. The extract, called podophyllin, has been used to treat complaints such as warts and is now used against cancer. Podophyllin seems to work by interrupting cell division and so it can be used to stop the multiplication of abnormal cells in a tumour. The active compound in the crude extract is called podophyllotoxin, but it has harmful effects on normal cells. The podophyllotoxin can be modified chemically to make etoposide which has fewer side effects. It would normally be much too expensive to make etoposide artificially, but the supply of wild plants is limited – all but two species of the plant are rare and all of them are slow growing; even sowing seed has to be done straight from the fruit or they will not germinate. Propagation by cuttings is also slow, so research has been done on producing new plantlets by taking root explants from seedlings and growing them in tissue culture. These can then be potted on and grown.

Micropropagation enables large numbers of identical plants to be produced quickly.

A single horticultural company can produce hundreds of thousands of plantlets a week without the need for huge areas of land. The companies may have to install expensive equipment, pay attention to sterility and develop special staff skills, but conventional propagation is also demanding of equipment and hygiene. Cloning also allows far more infection-free plants to be raised per unit area of bench space than ordinary methods.

5.14 ALGAE

Processes that use algae, apart from traditional gathering for food, are still in the trial stages. They grow well in warm, shallow ponds with large surface areas but this is not economical industrially as land is expensive and so is warming the water. However, the main problem with using algae is getting sufficient light to the cells.

A system is being researched using algae in a photobioreactor. The algae are grown in tubes in nutrient solution and controlled warmth and light. The arrangement results in a large surface area to absorb light and more cells are grown than in ponds. The conditions can be manipulated to direct cell metabolism to synthesise the desired compound. Algal cells can be harvested from the tubes and the product extracted. It is hoped that a number of products could be made this way including β carotene which is widely used as a food colour, and polyunsaturated fatty acids which are used as dietary supplements.

QUESTIONS

5.21 What are the advantages of using immobilised plant cells to make flavours and pharmaceuticals?

5.22 Why can't plant cells be grown in fermenters like bacteria?

5.23 What are the advantages of using cloned plants instead of collecting seeds?

5.24 Re-read the passage on growing plant cells in Chapter 2, then outline the procedure you would use to produce a clone of violet plants for a mini-enterprise scheme in time for Mother's Day in March.

5.15 ANIMAL CELL PRODUCTS

Cultured animal cells have been used to produce vaccines since 1947 when foot-and-mouth vaccine was made from tissue culture. However, monolayers of cells in small containers are not suitable for obtaining large quantities of cells, so work has been done on a fermenter for animal cells, but as yet there is no equivalent to a bacterial fermenter. Cells can be grown in suspension cultures but they grow slowly and the medium quickly becomes unfavourable. Trials are being carried out on systems which immobilise cells on inert supports, much as enzymes are now used, and a number of useful cell lines have been shown to grow and secrete product when immobilised on a trial scale. If these methods are successful, maintaining the medium and extracting product will be much easier.

As techniques improve, the trickle of products has become a flood. Large scale antibody production has touched everyone's life in some way, so it is described in detail below. There are also other products that have been cleared by safety committees which are now coming onto the market. Amongst these are tissue plasminogen activator from Chinese hamster cells; blood clotting factors and erythropoetin are made from other cell lines.

The advantage of tissue culture is that the molecular structures are the same as in a whole animal so that no immune responses are provoked in the recipient. A major project which has swallowed up an enormous amount of money and taken years is that to produce interferon, an antiviral chemical produced by human cells. As yet there is still no real sign of large-scale therapeutic use against viruses but the systems for growing the cells which synthesise it have been developed.

5.16 MONOCLONAL ANTIBODIES

Antibodies are proteins made by a type of white blood cell called B-lymphocytes which do not grow well in culture. Antibodies can be extracted from blood serum, but they are found in small quantities and are a mixture of antibodies to a variety of antigens.

The development of cells that make a specific antibody and grow in large-scale culture came from research in a different area of biology. Animal cells in culture occasionally fuse together, particularly if certain viruses are present. It was discovered that a chemical polyethylene, glycol (PEG), could have the same effect, and the fused cells did not even have to come from the same species. Hybrids between species such as man and mouse are useful in genetic studies to locate certain genes. In 1975 Kohler and Milstein managed to make a line of cells from fused mouse spleen cells and cells from a tumour called a myeloma. The cell line was called a **hybridoma**. The cell line had the spleen cell ability to make antibodies, and the ability to grow in culture came from the tumour cell. At long last there was the possibility of making antibody that could be used in medicine in large quantities.

The process

A new cell line has to be made for each type of antibody. Antibodies are made in response to exposure to antigens found on cells of all sorts including bacteria. Mice are injected with the chosen antigen and some of their spleen cells become dedicated to making antibody to that antigen; these cells multiply. The spleen is removed and the cells are mixed with tumour cells from a mouse plasmacytoma and PEG. Some cells fuse and are separated out and cultured. Only a very small proportion of the fused cells are antibody-secreting cells. Unfortunately, not all these are dedicated to making the desired antibody – some of the cells will make antibodies to other antigens the mouse has met during its life. These have to be removed before the desired hybridoma can be grown on a large scale.

The hybridoma cells make most antibody if they are injected into a mouse to grow, but they can be cultured *in vitro* – selected cells are grown on soft agar where they multiply to form a clone synthesising antibody, hence the term **monoclonal antibody**. The antibody is secreted into the culture medium from which it is extracted. Alternatively, the hybridoma cells can be deepfrozen until they are needed. Unfortunately, the whole process is more difficult if human-specific antibody is needed, as human plasmacytomas are not very suitable for fusion.

Monoclonal antibody is made by fused cells and can only combine with one sort of molecule.

Monoclonal antibody be used in a variety of ways: the molecules can be linked to resin beads, enzyme or dye molecules and are stable enough to be used in commercial products. Monoclonal antibodies have been put to many uses including diagnosing and treating infections, locating molecules, tracking cancer cells and investigating foods.

Monoclonal antibodies in immunosorbant columns

Immunosorbants are used to extract materials such as proteins from a mixture to make a very pure product. Monoclonal antibodies are raised against the chemical to be purified. The antibodies will combine with this chemical but not others in a mixture. They are immobilised by linking them to inert beads then packed into columns or on a permeable framework. The process is illustrated in Fig 5.17. The columns can extract molecules from concentrations of less than 1 per cent and give a high quality product.

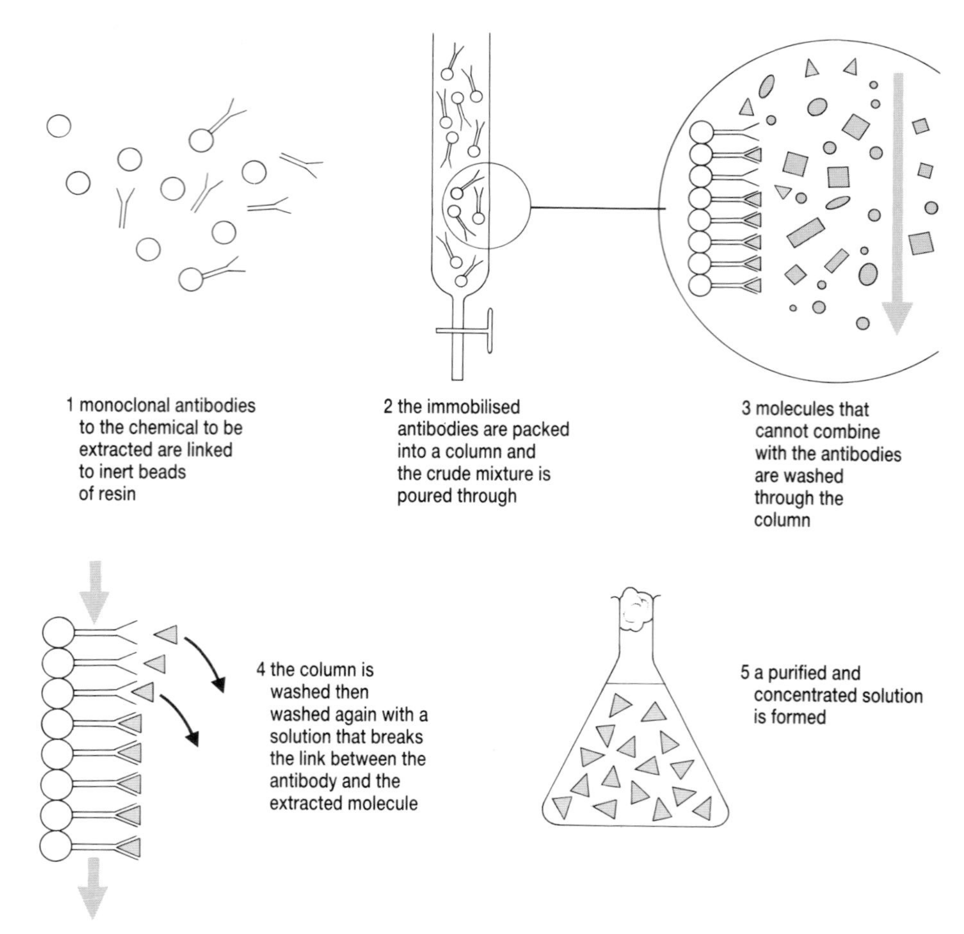

Fig 5.17 Monoclonal antibody immunosorbant columns.

Diagnostic testing

Many diagnostic kits use monoclonal antibodies. The antibody molecules in the kit are linked to other molecules, for example enzymes, which can bring about a colour reaction when the antibodies react with their specific antigen, even in very low concentrations. These kits are quite cheap compared to other diagnostic procedures and give results very quickly, from almost instantly to about 24 hours. For example, diagnosing infections is time-consuming; if a patient is ill with an infectious disease the treatment sometimes has to start before doctors can be sure of the cause. Usually samples are taken from the patient and plated out on different agars to check growth patterns; they are also tested against antisera for positive reactions. After two or three days the results of the tests are collated to confirm the diagnosis. Now there are kits available with antibodies to specific microbial antigens that reduce the time taken to less than 24 hours. Samples are mixed with a range of specific antibodies linked to dyes and incubated. A colour change happens if the particular antigen is present. A reference key of colour changes indicates what sort of organism is present.

Another enzyme and antibody system can also be used to detect molecules in very small quantities, known as **ELISA** which is Enzyme Linked Immunosorbant Assay. It is sensitive and can be used on very tiny samples, as small as $\frac{1}{1000}$ cm^3. Monoclonal antibodies to the substance being monitored are attached to an inert material and the solution to be monitored is passed over them. The antibody reacts with the target molecule. Enzyme linked to a different antibody is then added to the mixture and will attach through the second antibody to the complex already formed. The enzyme active site is not involved in the attachment and if it does not bind it is washed out. Next the enzyme substrate is added to the mixture and any bound enzyme reacts with its substrate to make a coloured product. The user sees a colour change take place if the chemical to be detected is present. It can be used for a variety of purposes including detecting the presence of drugs in the urine of athletes.

Fig 5.18 A pregnancy testing kit which uses monoclonal antibodies. The device is dipped into urine and a blue line of beads, trapped by antibody binding, forms in one of the windows telling the women whether she is pregnant or not.

Pregnancy testing

A pregnancy testing kit, shown in Fig 5.18, is a good example to illustrate 'do-it-yourself' diagnostics. A sign of pregnancy is human chorionic gonadotrophin (HCG) in a woman's urine. It is made after conception in small amounts at first, but the sensitivity of the kits allow it to be detected as soon as a period is missed, with great accuracy. The kit and the exact mechanism varies from manufacturer to manufacturer. In one, the test unit has a wick which leads into a chamber filled with coloured resin beads carrying monoclonal antibodies to part of the HCG molecule. The wick is placed in urine which passes up the wick and any HCG will combine with the antibody on the beads. The urine pushes the beads into another section where there is a membrane carrying another antibody. This antibody is combined to another part of the HCG molecule so that beads attach, via an HCG molecule, to the membrane. A window in the device allows the user to see the colour of the beads trapped on the membrane. If the woman is not pregnant the beads are not trapped but go on to another part where they attach to a third kind of antibody and can be seen there as a negative result.

QUESTIONS

5.25 What is meant by the term 'monoclonal antibody'?

5.26 What reasons could there be for worry about the safety of products made from animal or human cells which are used in humans?

5.27 Explain the advantage of using immunosorbant columns for extracting a chemical from a mixture.

5.28 Monoclonal antibodies attached to radioactive materials can be used to locate cancerous cells in the body. How would you locate these antibodies when they are attached to their target cell?

5.17 MINING WITH MICRO-ORGANISMS

All the processes so far are high technology processes using large plant, computer control, sophisticated equipment and an educated workforce. Not all processes are like this – using micro-organisms to extract minerals requires very little sophisticated machinery and even uses up scrap metal. The extraction can be done at the mine, which may be in a remote or underdeveloped area. The micro-organisms release metal ions from ore compounds and alter the physical environment so that metals precipitate in a form that can easily be refined.

Copper extraction

Copper is one of the most important industrial metals. Even low grade copper ores are worth mining as the demand is so great. The ores are known as porphyrics and contain 2–5 per cent copper. Conventional copper production involves three processes: extraction of the ore, smelting and refining the metal. The first two processes are done at the mine, if power is available – both use large quantities of coal. After extraction the ore is fed into a reverb furnace then into a Bessemer converter to drive off iron and sulphur impurities. This produces an impure metal which is sold relatively cheaply to the refiners. Refining is by electrolysis and is usually carried out near the consumer factories in developed countries. Costs are reduced if refining can be done at the mine as land value will be low, and in some major producer countries such as Chile and Zambia the labour costs are lower than in Europe. The mining process is expensive, energy consuming, polluting, dangerous and very unsightly. The raw materials are relatively cheap but the countries which mine the ore and sell it cheaply generally have to buy the refined copper back as expensive products.

The biology

Micro-organisms are used increasingly to extract copper; between 10 and 15 per cent of the world's copper is now extracted microbially. The technology is old but simple; even the Romans are thought to have used it. The organisms are chemoautotrophs able to use low grade ores. They can even extract copper from the waste heaps of ordinary copper mines. The usual ore is chalcolite which is copper sulphide (Cu_2S), and the micro-organism is *Thiobacillus ferrooxidans,* a bacterium which carries out in-organic oxidations to gain energy for metabolism and in doing so creates conditions that release copper in a soluble form. In damp conditions the organism oxidises sulphides such as iron(II) sulphide (pyrites) to iron(III) sulphate – an energy-releasing reaction. Sulphuric acid is made at the same time, making the environment very acidic. Both of these products play a part in the extraction of copper from copper sulphide as copper sulphate.

The process

The process is illustrated in Fig 5.19. The crushed ore or waste from copper mines is heaped up on a waterproof surface. It is sprayed with water with some sulphuric acid to start the process, though once it is going drainage water from the dump can be recycled. The acid conditions encourage the growth of the micro-organisms already present in the soil and ore at the mine at low levels. These organisms oxidise the ore and also oxidise iron pyrites impurities, releasing iron(III) ions. The copper compound made from the ore reacts chemically with the iron(III) ions, releasing copper. At the same time sulphate ions are formed which react with copper to make copper sulphate which drains out of the dump into a shallow pond. Scrap iron thrown in the pond displaces the copper from copper sulphate and copper metal precipitates out. The precipitated copper is scraped out and refined electrolytically. Alternatively the copper can be extracted and re-fined directly by electrolysis. A large proportion of the copper is extracted in this way. The process takes up a lot of space and is ugly to look at, but it does not require high technology, furnaces and large amounts of fuel.

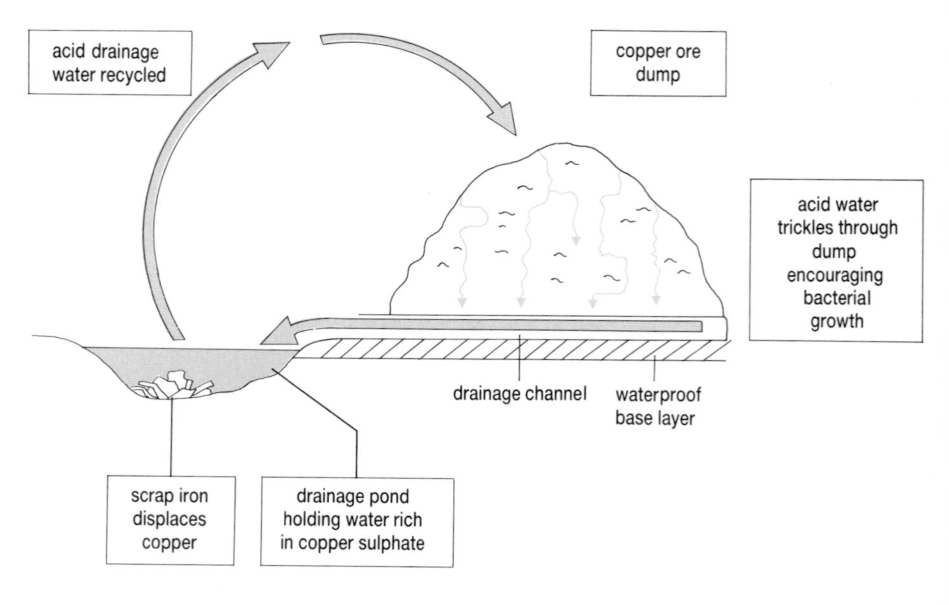

Fig 5.19 Copper mining with micro-organisms.

Other extractions

A large amount of uranium is extracted using microorganisms in the USA even though it is not economic. Most of the world's uranium for civil and military nuclear power comes from Namibia or the USSR with smaller deposits elsewhere, so its supply is a sensitive issue. The process is very similar to the copper extraction outlined above. *T. ferrooxidans* causes the release of iron(III) ions from pyrites in uranium ore, and these oxidise the uranium ore to uranyl sulphate. Uranium is then extracted from the uranyl sulphate. Potentially other valuable metals could be extracted in a similar way.

Micro-organisms could even be used to extract oil from underground reserves, oil shales and tar sands. Some research has been done in this area. Organisms could, for example, be pumped into the reserves in a flow of water. There they would degrade the crude oil, and petroleum derivatives could be pumped back up to the surface or collected by seepage. The advantage is that less plant would be needed to make a useful product. One microbial product is already used to extract oil: water pumped into an underground reserve to push out the oil does the job better if it has xanthan gum, made by *Xanthomonas*, added to it because the water becomes thicker and more effective at removing the oil from the rocks.

QUESTIONS

5.29 Why can mining copper with micro-organisms be a cheaper process than the conventional method?

5.30 Discuss the contribution of the use of micro-organisms to helping the world population overcome the problem of diminishing natural resources. You are free to write on any aspect of this topic, but you may wish to refer to food and fuel products and the recycling of scarce resources. *(JMB)*

5.31 **(a)** Describe how micro-organisms play a part in the synthesis of a range of products useful to man.
(b) Give an account of the commercial preparation and extraction of any one such product. *(JMB)*

SUMMARY

Micro-organisms and cells can be grown on a large scale in fermenters. They produce a wide range of products with various metabolic capabilities. Fermenters are designed to give the inhabitants the optimum growth conditions for the synthesis of the product, though changing from laboratory scale to factory scale does introduce new problems.

The products made by micro-organisms include enzymes, primary metabolites and secondary metabolites. Whole cells may be used to make a product, or they may themselves be the product. Metabolites made by cells are extracted and processed into a product. Animal and plant cells are also used successfully to make products.

Not all microbial processing uses high technology.

Theme 3

MICRO-ORGANISMS AND DISEASE

I think great things are coming to pass. Joseph Meister has just left the laboratory. The last three inoculations have left some pink marks under the skin, . . . we approach the final inoculation, which will take place on Thursday, July 16th, . . . The lad is very well this morning, and has slept well, though slightly restless; he has a good appetite and no feverishness . . . Perhaps one of the great medical facts of the century is going to take place; you would regret not having seen it.

Louis Pasteur writing about the first human test of his anti-rabies vaccine, used to treat a small boy who had been bitten by a rabid dog, 1885.

Discoveries about the nature of infectious disease in plants and animals, and how plants and animals protect themselves against pathogens, has led to the development of therapeutic chemicals which can be used to treat infections. Other techniques such as spraying crops with protective chemicals, vaccinating people and improved hygiene have reduced the incidence of infectious disease.

Prerequisites

Read sections 1.6 to 1.13 which describe the biology of disease-causing micro-organisms. It is helpful to have studied blood and the immune system at A-level, and have some knowledge of plant structure.

Chapter 6

THE FIGHT AGAINST DISEASE

LEARNING OBJECTIVES

After studying this chapter you should be able to:

1. understand how organisms cause disease;
2. explain how diseases are transmitted;
3. explain the normal defence mechanisms in animals and plants;
4. describe the defence mechanisms which are induced as a result of infection;
5. describe how natural defensive mechanisms can be enhanced;
6. appreciate how therapeutic agents can be used to treat and control disease.

6.1 THE CAUSES OF DISEASE

The word 'bacteria' automatically makes a lot of people think of disease. Micro-organisms do cause disease, but there are other causes too, for example illnesses can arise from malfunctions of the body such as diabetes, or from nutritional deficiencies. Yet still 'micro-organisms' and 'disease' remain inextricably linked. This is justified with viruses because they are intracellular parasites which almost always disrupt normal cell activity, but not all bacteria are harmful – in fact most species have very little effect on our wellbeing. Those bacteria which do affect us play an important part in human affairs. We have only to think of the economic changes brought about by plague in Europe in the Middle Ages, or the constant drain on resources caused by diseases such as cholera or malaria in many parts of the world today.

Fungi are not usually included in our catalogue of 'nasties' – we don't see headlines like those in Fig 6.1! – yet many animal diseases are caused by fungi, though these tend to be trivial infections unless the sufferer is debilitated for another reason. For plants, things are very different – bacteria cause few problems but fungi are a major cause of disease.

All the main groups of micro-organisms – bacteria, fungi, viruses and protozoa, have disease causing members. Plants also have a few algal parasites but as a group they are not disease-causing. The most important disease-causing agents today are the viruses. There are few therapies for an animal or plant suffering from a viral disease: virus diseases have to be prevented rather than cured. Protozoal diseases also pose problems because of the large number of people who suffer them, and because it is difficult to control the transfer of the protozoan from one victim to another.

Each day animals and plants come into contact with hundreds of harmful micro-organisms, but only a few of these encounters lead to disease. All living things have a range of defences which prevent disease-causing micro-organisms from gaining a foothold. Occasionally, however, an organism will establish itself and cause problems. Some diseases, such as septic infections, are less important now than in the past because there are treatments which will cure them.

Fig 6.1 We don't have to worry about headlines like these!

A pathogen is a micro-organism that causes disease. Its ability to cause disease is its virulence.

A disease which can be passed from one sufferer to another is described as an **infectious** or **communicable** disease. Organisms that can cause disease are described as **inductive agents**. A species which causes a disease when it infects is a **pathogenic** species, however anyone who has had a cold and has also had influenza knows that not all pathogens cause equally severe diseases. Neither are disease-causing organisms equally capable of causing disease. The ability to cause disease is called **virulence**. Some organisms, such as the leprosy bacterium or the food poisoning bacteria, have low virulence. It can take many encounters or large numbers of organisms to cause the disease. Other diseases such as smallpox are highly virulent; only a few organisms are needed to start the disease. A strain which cannot cause a disease is described as **avirulent**.

Epidemics

In any population of plants or animals there will always be some individuals suffering from an infection. If a few individuals in a small area suffer from a disease we talk about an **outbreak** of disease. If a large number of individuals in several communities suffer from the same infection it is an **epidemic**, and when very large numbers in different countries suffer then it is a **pandemic**. These terms should only be used to describe infections affecting people but sometimes they are used for infections of plants and animals. The most recent human pandemics have been influenza pandemics. It is said that 'flu killed more people in the 1918 pandemic than were killed in the First World War.

Epidemiology is the study of the spread of disease and the factors that influence spread. If enough is known about the spread of a disease measures can be introduced to control, or even eradicate, a disease from an area. Initially an epidemiologist will try to find the sources of an infection and the mechanisms of transmission. Once these are established then health authorities and other bodies can focus their efforts in the most effective ways.

A pathogenic species of micro-organism is only found where there are enough potential victims to keep the micro-organism population going and where transmission from one victim to another is relatively easy. For example, at one time measles virus was unknown on many islands in the Pacific as there were not enough people to keep the virus circulating, and so people had no immunity to it. When European travellers brought measles to the islands, it was able to infect a large proportion of the population. A pathogen which is normally present in an area is said to be **endemic**. Though a disease may be endemic, it does not necessarily mean that the population suffers a great deal from it. For instance, plague is found in wild animals in North America, but it is seldom transmitted to people and few suffer from the disease.

An endemic disease is one which is native to an area.

6.2 HOW MICRO-ORGANISMS CAUSE DISEASE

Pathogens are not perfect parasites: their activity may disrupt the host so much that its vitality is affected, threatening the host's life. Some pathogens destroy tissues and release chemicals which influence the host's metabolism. Viruses cause cell damage because they interfere with normal cell functions and because many leave the host cells by breaking open the external membranes. Other pathogens interfere with protein synthesis in cells, or with transport across cell membranes. Micro-organisms may release enzymes that degrade tissues causing damage and **necrotic** areas of dead cells like those shown in Fig 6.2. Victims may also suffer as a consequence of their own defence mechanisms: for example, immune complexes between antibodies and micro-organisms are made in diseases such as leptospirosis. These are big enough to lodge in small capillaries in the liver and kidneys, blocking the passage of blood to tissues.

Fig 6.2 An infection with *Ascochyta* fungus has caused these blackened necrotic areas on the bean leaf.

Toxins

Toxins are chemicals made by pathogens which have harmful effects on the host. The effects are felt even in uninfected tissues as the toxin is transported through the host's tissues from pathogens lodged elsewhere. **Exotoxins** are soluble compounds secreted by the organisms into their immediate environment; **endotoxins** are components of the outer cell membranes of Gram negative bacteria.

Microbial toxins made in animals include many of the most potent poisons known. The tetanus toxin made by *Clostridium tetani* is an exotoxin which binds to nerve endings and blocks nerve impulses so that the muscles of the body remain contracted in a severe spasm called tetany. The old name for the disease was lockjaw, which describes vividly its effects on the facial muscles, as shown in Fig 6.3. The **enterotoxin** made by cholera organisms is an exotoxin which is active in the gut. It affects the cells of the gut wall causing them to pump out valuable electrolytes and fluids. The victim dies of dehydration. Endotoxins cause fever and fluid loss.

Toxins are not made exclusively by bacteria; fungi make them too. In Chapter 3 there is an account of the toxin aflatoxin produced by the fungus *Aspergillus flavus* which binds with host DNA, blocking RNA synthesis and the expression of the host's genes. Some plant pathogens make toxins but there are few which have been shown to definitely produce symptoms in the absence of the pathogen.

Fig 6.3 A contemporary drawing of a soldier in the Napoleonic Wars suffering from tetanus. The tetanus toxin causes muscles to stay contracted, including the respiratory muscles. One of the few treatments available at the time involved knocking the patient's teeth out and inserting a tube into the patient's lungs to ventilate them.

Immune system damage

Many pathogens directly affect the cells of animal immune systems reducing the ability of the host to overcome the disease. Some bacterial toxins affect the phagocytic cells; they may kill them outright or inhibit them so that they do not take up the invading organisms. Others cause blood clotting round the infection which bars the migration of white cells to the site of infection. A particularly severe disease affecting immune system activity is AIDS, acquired immunodeficiency syndrome, in which a virus infects a type of white cell called a T-helper cell and prevents it from carrying out its role in antibody production.

Some pathogens are able to disable host defences by a variety of mechanisms.

Sometimes damage is caused by the response to the infection. Some pathogens, especially viruses, have antigens which protrude from the surface of host cells and provoke immune system cells to react against them. For example leprosy bacilli in humans grow very slowly in nerve cells and eventually cause a loss of function. The bacteria cell walls are extremely tough so components remain in the cells even after the bacterium is dead. When the body's defence mechanisms detect the bacteria or their components, white cells respond and damage the nerve cells in the process.

Leaf damage

It often seems that an ailing plant has only one symptom – its leaves go yellow and it dies – irrespective of what it is suffering from. However there are more responses than this: an infection can cause alterations in plant physiology which cause infected leaves to retain materials made in photosynthesis that would normally be translocated to other parts of the plant, depriving the shoots, seeds and roots and reducing their development. The pathogen may also cause the import of materials from other parts of the plant. The two processes can make infected leaves appear healthy and vigorous right up until the plant dies. The plant may show few other symptoms than this change in the pattern of storage of carbohydrate until the infection has weakened the plant enough for other symptoms to show. Sometimes the tissues next to the infected tissues increase their activity, making patches which are described as 'green islands'.

Fig 6.4 Virus infections often block chlorophyll synthesis. An infection with cucumber mosaic virus has resulted in patches of cells being paler than those around them. Many virus infections, as in this plant, first show in the cells closest to the transporting tissues.

Other activities may change in infected cells, for example the respiration rate and membrane permeability may increase. It is thought that this might fuel defence mechanisms, though this may well be to the pathogen's advantage. Root-invading fungi interfere with the uptake and transport of water and minerals. Viruses may block chlorophyll synthesis, so cells are paler and leaves have lighter blotches and photosynthesis is reduced. We do not know why a virus causes pale yellow blotches or a **mosaic**, as shown in Fig 6.4, in one species of plant but causes streaks down the leaves in another species. Some effects are thought to be caused by alterations in the balance of plant growth regulators since these, or their analogues, are secreted by some pathogens, causing symptoms such as leaf curls.

QUESTIONS

6.1 What is the difference between an epidemic and a pandemic?

6.2 What is meant by a virulent strain of bacteria?

6.3 Briefly describe two ways that micro-organisms can cause damage in
(i) plants and
(ii) animals

6.4 What is a toxin?

6.5 Describe the action of one toxin.

6.6 Why do nerve cells suffer damage in a leprosy infection?

6.7 An investigation of red spot disease, a fungal disease in a particular plant species, was carried out. The plants had all been grown from the same batch of seed in identical conditions. Two of them were inoculated with spores of the disease causing fungus. Each was illuminated and fed with radioactive labelled carbon dioxide ($^{14}CO_2$) through one leaf, then the distribution of the labelled carbon was monitored six hours later. Fig 6.5 shows the distribution of radioactivity.
(a) How does the fungus affect the distribution of material made in photosynthesis?
(b) Plant B acts as a control for both infected plants. What would have been a better control for plant C?

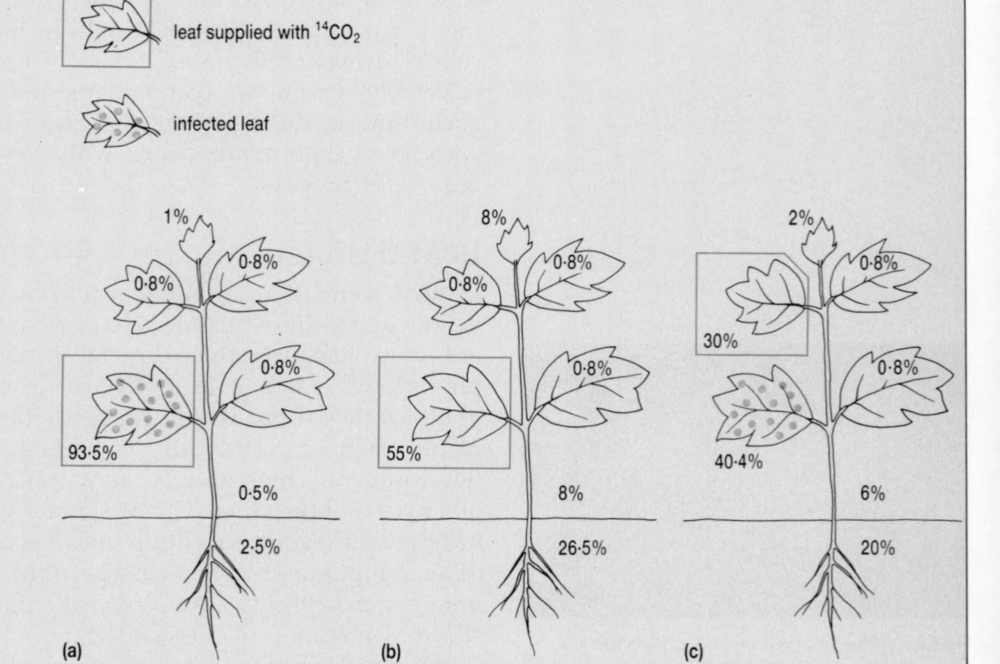

Fig 6.5 Distribution of radioactivity in a healthy plant and diseased plants.

6.3 HOW MICRO-ORGANISMS ARE TRANSMITTED

Animal and plant pathogens come from a variety of sources. Some are those found normally on skin, leaf surfaces, mucous membranes and around roots which have unexpectedly gained entry to the interior. Many more come from other infected individuals. Micro-organisms are passed from one victim to another in a number of ways. Those are not motile have to rely on passive transmission from one individual to another. Others exploit **vectors**, such as insects, to carry them about. Motile micro-organisms may able to make their own way to a potential host, but the distances they may have to travel are huge compared to their size. The transfer time is dangerous for many pathogens: the temperatures may be lower than their optimum if they are adapted to life inside mammals and birds, and they are subject to drying and UV light. The length of time that an organism can survive outside its host varies: those passed by direct contact have a very short survival time but others may be transmitted as resistant structures such as spores.

Direct contact

Diseases can be transmitted by direct contact from one infected individual to another. Often the contact points in animals are the mucous membranes, which are thinner and softer than most outer surfaces, though infection through the skin is a means of transmission for several organisms.

Directly into blood

This is an uncommon method of transfer, but the few diseases which are spread this way are severe. Transfer occurs when blood or body secretions from one individual enter the bloodstream of another. Usually this happens when a secretion such as seminal fluid or saliva comes into contact with a damaged mucous membrane allowing entry directly into capillaries. Pregnant women may pass infections across the placenta to the developing embryo. Sexual activity is a frequent means of transfer, as are infected hypodermic needles. Some remarkable instances have been recorded of other events leading to transfer, such as communal bathing or activities which have led several people to scratch themselves on the same thorn or splinter of wood, but these are exceptional.

Fig 6.6 A combination of damage caused by birds pecking the fruit, falling to the ground, and slug activity afterwards has left openings in the protective skins of these apples, which have provided an entrance for fungi.

Wounds and bites

Organisms which enter through broken skin or leaf surfaces are often opportunistic. They are normally confined to the surface of a plant or animal, but a wound in the protective surface caused by an accident (like the apple in Fig 6.6), wind damage, bites, the ravages of chewing insects or browsing herbivores provides the entry point for pathogens.

Air- and soil-borne transmission

Organisms can be transmitted in air currents: most fungi, disease-causing or not, rely on air currents or rain splashes to disperse their spores. Organisms that infect animal respiratory tracts are also spread by exhaled air or by droplets of mucus spreads by coughs, sneezes or running noses. When someone sneezes, a fine spray of droplets of saliva and mucus is produced which spreads through air currents. Large droplets settle quickly but smaller ones remain suspended until the moisture evaporates leaving small particles called droplet nuclei. These take a long time to settle and are easily inhaled by another person.

Plants suffer particularly from soil-borne diseases. The causative agent, either dormant or growing in the soil, waits until a suitable plant grows nearby then it infects.

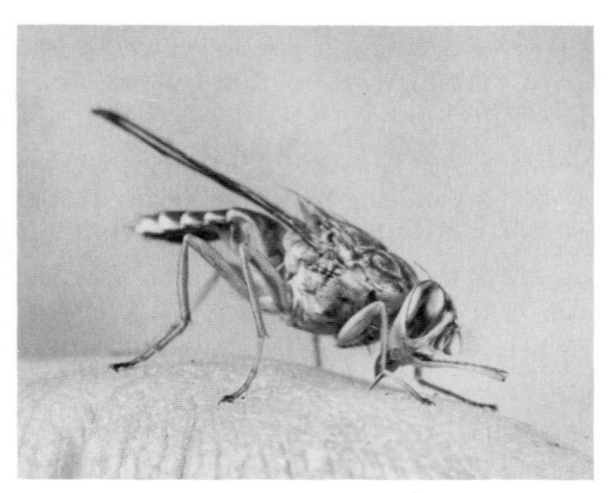

Fig 6.7 The tsetse fly, *Glossina*, feeds on blood and may carry parasitic protozoa in its salivary glands. The parasites are *Trypanosomas* species, which cause sleeping sickness. The parasite's presence in the feeding parts can interfere with the insect's ability to monitor blood flow while feeding; causing it to feed more frequently, and for longer, assisting the parasite's chances of transfer.

Water- and food-borne transmission

Pathogens get into food with the ingredients; from the activity of other organisms such as houseflies which transfer them from other places; and as a result of human failure to observe basic rules of hygiene such as washing hands after visiting the toilet. Enteric diseases are caused by a variety of organisms, including typhoid and cholera bacteria found in the faeces of infected people and in sewage-contaminated drinking water. Many raw meats and poultry carry large numbers of salmonella bacteria, which also cause a form of gastroenteritis, that are transmitted by handling and undercooking.

Vectors

Many viral and protozoan pathogens are transmitted by vectors. The most important vectors are the insects. Biting flies, bugs and aphids transfer pathogens as they feed on blood or plant sap. The peach-potato aphid, *Myzus persicae*, has been recorded as carrying about 70 different plant viruses. Mosquitoes and aphids have mouthparts adapted as stylets that penetrate through skin and between cells, as can be seen in Fig 6.7. Salivary secretions which help in feeding, carry the pathogen into the victim.

Vectors may transmit pathogens mechanically or the pathogen may be biologically linked to its vector.

Some micro-organisms are merely carried mechanically on the mouthparts from one individual to another, but many have a close relationship with their vector. The pathogen may live for part of its life cycle in the vector, or may remain in it permanently. Vector-carried pathogens have to be adapted to survive in two different hosts, for example the plague bacterium is found in rats and humans but is carried by a flea. Similarly the malaria parasite is adapted for survival in human livers and bloodstreams, yet is also capable of surviving an *Anopheles* mosquito's digestive system and can multiply in the insect's equivalent to a bloodstream, the haemolymph. Many vector-carried organisms never have to face a hostile environment as they are always inside a suitable host.

Table 6.1 Selected diseases and their means of transmission

Disease	Inductive agent	Transmission
athlete's foot	*Trichophyton rubrum* (F)	soil- and water-borne spores
apple fruit rot	*Penicillium/Rhizopus* (F)	wound entry by spores
leptospirosis (Weil's disease)	*Leptospira interogans* (B)	wound entry from contaminated water, infects man and rodents
measles	measles virus (V)	inhalation of virus
polio	poliomyelitis virus (V)	drinking water
PVX disease	potato virux X (V)	direct contact with infection
sleeping sickness	trypanosome (P)	vector, bite by tsetse fly
syphilis	*Treponema pallidum* (B)	direct contact in sexual activity
wheat rust	*Puccinia graminis* (F)	air-borne spores infect cereals

Note: B=bacterium; F=fungus; P=protozoa; V=virus

QUESTIONS

6.8 Give one example of a pathogen from each of the main groups of disease-causing micro-organisms and its means of transmission.

6.9 Why do some organisms need a vector to transfer them?

6.10 Name two vectors and the pathogens they are associated with.

6.4 DEFENCE MECHANISMS

As micro-organisms are so abundant, animals and plants would be expected to suffer from more infections than they do. However animals and plants have co-existed with micro-organisms for millions of years and have evolved defence mechanisms to repel the micro-organisms which infect them. Animals and plants have defensive strategies which make it difficult for pathogens to gain entry, backed up by a range of defences which operate against any invading micro-organisms, and animals additionally have specific defences against individual sorts of pathogens.

Surprisingly, one of the best protections against pathogens is the presence of other micro-organisms. The surfaces of plants and animal skin, gut and respiratory tracts carry a host of micro-organisms called **commensals**, which are harmless as long as they are confined to the surface. Commensals are adapted for life on outer surfaces and colonise successfully the available ecological niches. Any incoming pathogens, which are usually better adapted for life inside the host than outside, have first to compete with the established commensals. It is difficult therefore for pathogens to establish an initial colony and spread unless circumstances are particularly favourable. Commensals are kept in check by the host's normal defence mechanisms. If these fail, for example, if the surface tissues are damaged or an animal's immune system is suppressed, commensals may enter and cause **opportunist infections**. Plants suffer fungal opportunist infections when they are damaged by herbivores or severe weather. Doctors treating transplant patients and AIDS sufferers have to be alert to opportunist infections by skin commensals.

Commensals reduce the likelihood of a pathogen becoming established.

6.5 NON-SPECIFIC ANIMAL DEFENCES

Natural barriers and chemicals

The skin

Pre-existing barriers to microbial entry and growth make it difficult for a pathogen to invade. There are a few natural entry points – sweat and sebaceous glands and entrances and exits to organ systems which are lined with mucous membranes. The skin is an arid environment to live in and a tough barrier to penetrate. The outer surface is an unstable layer of hardened cells with little moisture – a very difficult habitat for micro-organisms. Most of the micro-organisms on the skin cluster around the sweat and sebaceous glands, shown in Fig 6.8, where there are organic molecules and moisture, though there are chemicals in the secretions which inhibit microbial growth. If skin is injured by wounds or burns then micro-organisms can enter and cause infections, but clotting blood quickly makes a barrier and white cells are drawn to the wound, scavenging infecting organisms.

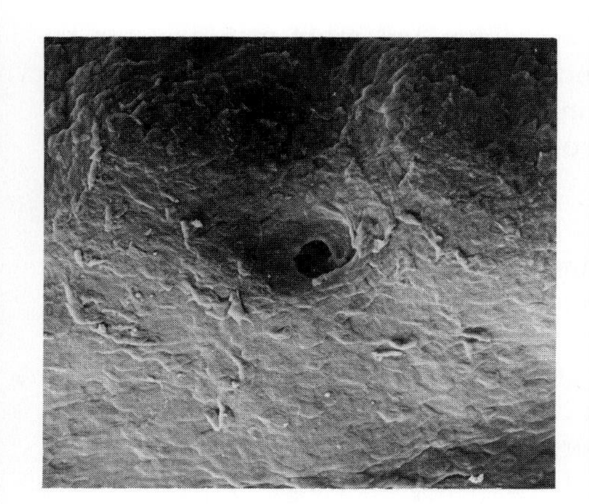

Fig 6.8 The entrance to a sweat gland is an oasis for microbial life; it is a source of moisture and nutrients as well as a way into the interior of the body. Further away keratinised outer cells make a tough, arid, and unstable surface as they flake away.

Entry points

The eye surface and mucous membranes such as those in the nose are suitable for microbial growth as they are moist, richer in nutrients and not hardened. The ways in which entry points such as the membranes, gut and respiratory passages are protected against pathogens in air, food and water are shown in Table 6.2. When people have a low stomach acidity, for example because of the action of certain drugs, they are much more likely to suffer from diarrhoeal infections since the conditions are more favourable to a wider range of micro-organisms. In general, the colon is more hospitable than the rest of the gut, but pathogens find it difficult to compete with the large numbers of anaerobic and microaerophilic commensals which live there and which are adapted to the low oxygen levels.

Table 6.2 Non-specific defences

Area	Protection	
skin surface	keratin	hardens cells
	fatty acids and lactic acid	inhibit microbial growth
mucus	IgA antibody	in tears, saliva and other secretions binds to micro-organisms
	lysozyme	degrades Gram negative bacteria walls
	mucus	traps particulate matter and passes to stomach
stomach	hydrochloric acids	kills pathogens
	proteases	degrade micro-organisms
small intestine	bile and proteases	degrade protein and emulsify lipid components
blood	transferrin	iron is sequestrated and levels reduced to below those needed for bacterial growth

Inflammation

Inflammation is a protective response to any damage or wound, not just damage caused by pathogens. The sequence of events brings together many different factors which hinder any microbial infection. Typical signs of inflammation are that the area is redder than usual, tender, swollen and warmer to touch. The symptoms are caused by more blood flowing to the area and the release of fluid from the bloodstream to the tissues. The capillary walls in the damaged area become more permeable and more materials can leave the bloodstream. Kinins are converted to bradykinin, which dilates the capillaries and keeps them permeable, and histamine, prostaglandins and other chemicals are released in the damaged tissues which activate white cells. As the blood vessels are now wider, the rate that blood flows through the tissues slows down and more materials can be exchanged between the plasma and the tissues. White blood cells can migrate through the capillary walls into the damaged tissues, drawn by the release of chemicals from a complex of plasma proteins called **complement**.

Complement

Complement is a collection of nine proteins in the blood that helps resistance to infection. They are secreted by a type of white cell called a monocyte and by cells in the liver as inactive proteins. Activation is a complicated process, but once activated, the proteins have many different functions. One type, C5a, acts as a chemotactic attractant for white cells. Another sort, C3b, sticks to micro-organisms making a recognition signal for phagocytes which engulf them. Other complement components affect the outer surfaces of microorganisms and help their destruction by making pores in the surface. The pores allow the contents to leak away or the cell to lyse. Mast cells, another type of white cell, are triggered by complement proteins to release the chemicals which enhance inflammation and attract more phagocytes to the area of inflammation.

Interferons

Interferons belong to a group of glycoproteins known as lymphokines and are made by a type of lymphocyte during virus infections. They are a non-specific protection against viruses, working while antibodies are secreted. So far three different sorts have been identified in humans, and other types have been found in animals, but they are only effective within their own species.

Interferons in tissue fluids prevent uninfected cells from becoming infected. They bind to the membranes of uninfected cells and stimulate them to make enzymes which degrade viral RNA and so block virus multiplication. They are not much use, though, if the infection is already well established. Interferons are responsible for many of the unpleasant effects of virus infections such as 'flu. Aches, shivers, tiredness and fevers are the typical symptoms, though other lymphokines can produce these symptoms too.

At one time interferon was thought to be the answer to viral infections but unfortunately it is extremely difficult to use as a therapeutic against viruses as it needs to be administered at the same time as the virus is starting the infection, which is often before any signs of the infection have become apparent. It is useful, though, in treating some cancers.

Interleukins are another sort of lymphokine which affect the immune system and may be incorporated into vaccines to improve the response to infections. Interleukin 2 stimulates lymphocyte growth in tissue culture and turns them into 'killer' or 'scavenger' cells, but more work has to be done before it can be put to trial in animals.

QUESTIONS

6.11 Which are the main entry points for micro-organisms into a human body?

6.12 Describe how the skin acts as a barrier to the entry of pathogens.

6.13 Explain how the presence of micro-organisms on the skin can be beneficial in the fight against disease.

6.14 Why does increased permeability of capillary walls near the site of the infection help overcome infections?

6.15 Name two compounds that draw white cells to the site of an infection.

6.16 Give one advantage and one disadvantage of using interferon to treat a virus infection.

6.17 Construct a simple table of the chemicals released during a microbial infection and their functions.

6.6 THE IMMUNE RESPONSE IN ANIMALS

Any activity which happens in the body as a result of encountering a foreign organism, or its products, is an immune response. All animals can produce an immune response, but some are more complicated than others. The human immune system is highly complex and the activity of the many different parts, their roles and interactions are still being investigated.

It has been known for a long time that people who contract severe infectious diseases and survive seldom suffer from them again. Both the Chinese and the Turks deliberately gave infections to people to protect them as early as the tenth century. In the eighteenth century Dr Edward Jenner exploited this observation when he used a cowpox infection to protect against smallpox. This process soon became widespread, and the term 'vaccination' was coined to describe it. An early vaccination can be seen in Fig 6.9.

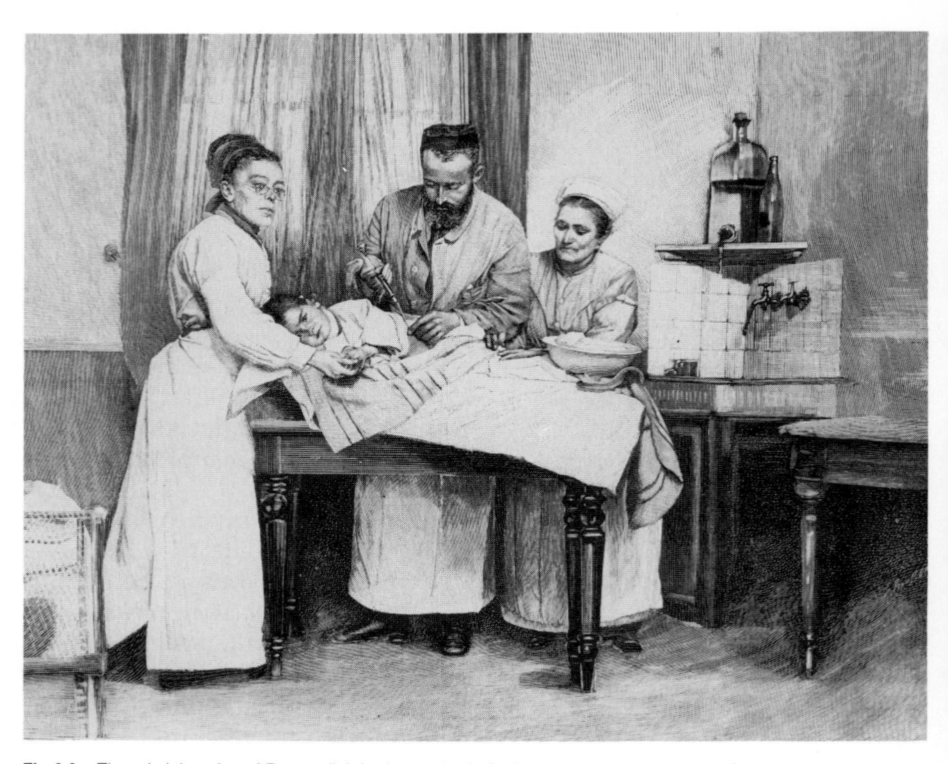

Fig 6.9 The administration of Roux's diphtheria vaccine in Paris last century. Lister's and Pasteur's ideas on hygiene and disinfection were now more widely practised in consulting rooms; this room represents 'state of the art' medicine in 1895. The medical staff are not in their outdoor clothing and have a sink with disinfectant; they also have clean uniforms, floors, implements, bedding and swabs, though they still have an ordinary table to work on.

When the blood of people who have recovered from an infectious disease is examined it is found to have much higher levels of proteins called **antibodies** than people who have not been exposed to the infection. They also have larger numbers of circulating white cells called **lymphocytes**. An invading pathogen carries a range of proteins, lipo-polysaccharides and other molecules in its outer layers. If these chemicals do not normally occur in the host's body they are recognised as foreign and an immune response is provoked. The foreign substances are called **antigens**. Some are small chemical groups forming part of a molecule or they can be whole molecules. The toxins and waste metabolites secreted by pathogens into their environment can act as antigens too.

Antibodies

Antibodies belong to a class of proteins called immunoglobulins secreted by lymphocytes in response to antigens in the body. When they are released into the blood they are known as antibodies. There are many different types of antibodies, and during an infection there is a massive increase in just one or two of these. Each different strain of pathogen provokes the production of slightly different antibody molecules which can only combine with that pathogen's antigens and no others. The ability of an antibody molecule to combine with only one antigen is called its **specificity**. The antibodies made during an infection can last in the bloodstream for months, long after the infection is beaten. A memory is retained within the immune system, so whenever more micro-organisms of the same type re-enter the body, even years later, a huge amount of specific antibody is made.

Antibody molecules are secreted by a type of lymphocyte called a **B-cell**. Each B-cell secretes antibody to one particular antigen and is described as being **dedicated** to that antigen. If a B-cell has antibody fragments matching a particular antigen, it is selected and multiplies quickly producing a clone of plasma cells that can produce large amounts of antibody. Within a few days of meeting a new antigen, high levels, or **titres**, of antibody can be detected, though production is slow at first.

The structure of an antibody molecule is shown in Fig 6.10. The specificity of the antibody molecule lies in the two ends of the arms of the Y shape. These sections are unique to each type of antibody and match exactly the surface antigens of the micro-organisms they were raised against. There is still speculation as to how so many different arrangements on the ends of the arms can be generated in such a short time.

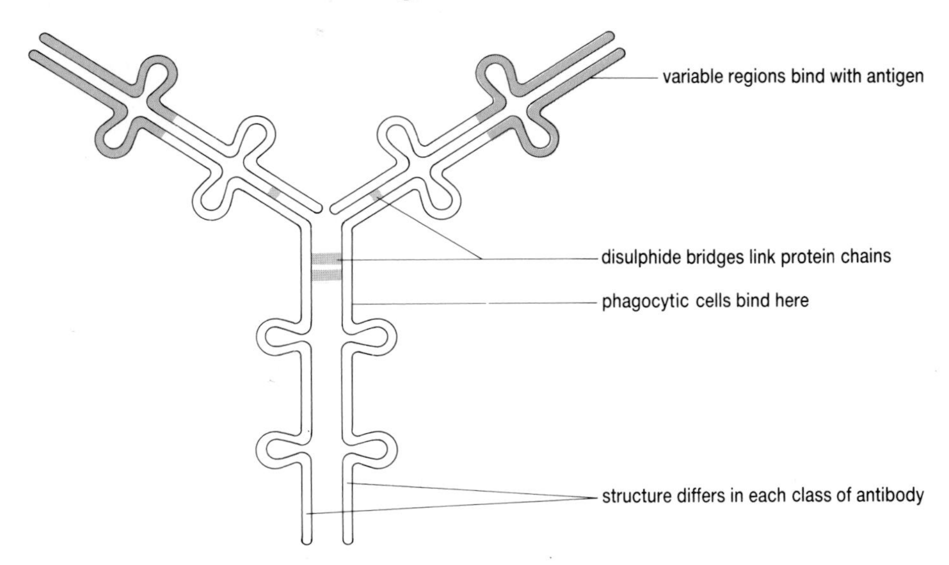

Fig 6.10 Structure of antibody molecule.

Different kinds of antibody molecules have different functions and are known by the names IgG, IgA and so on. Their roles are summarised in Table 6.3. Antibodies combine with antigens found in the capsules or cell walls of bacteria, with antigenic proteins in virus coats and with membrane antigens in protozoa. Some antibody molecules clump pathogen cells together or coat them, a process called **opsonisation**. These molecules then link to receptors on white cells called phagocytes and stimulate them into ingesting the coated pathogen. As each type of antigen provokes a different antibody, an invading bacterium with several surface antigens may have several different types of antibody molecules attached to it. Antibodies are also made to some toxins, for example those released in diphtheria and tetanus infections, which they neutralise by binding to the toxin molecules so that they cannot bind to host cells.

Table 6.3 The roles of antibody types

Type of molecule	Function
IgA	protects entry points. Common in secretions such as tears, on mucus membranes, in colostrum
IgD	bound to B-cells, involved in maturation of B-cells
IgE	linked to macrophages and mast cells, mediates release of histamine. Produced in allergic reactions such as hay fever.
IgG	most common antibody released into bloodstream by B-cells. Passes across placenta to foetus.
IgM	exists as a unit of five linked molecules in blood, made by B-cells

Antibodies are specific molecules. They combine with only one type of antigen molecule.

There are other sorts of antibody which are involved in general protection against infection. IgA on mucosal surfaces reacts with organisms found in the gut and airways; IgG is passed across the placenta to protect newborn babies. Both IgG and IgM play a part in activating the complement proteins. There is a rare condition known as hypogamm-globulinaemia in which the sufferer is unable to secrete any antibodies, and is therefore very prone to infections.

White cells

As well as antibodies, there are specific cells which play a role in the response to infection. Two main kinds of cells are involved in combating microbial infections; these are phagocytic cells which engulf particles carrying foreign antigens, and lymphocytes which are responsible for **cell-mediated immunity**. The different types of white cells differ in structure, as can be seen in Fig 6.11. There are also white cells called eosinophils in circulation which release chemicals into their immediate environment. Eosinophils affect large foreign bodies such as parasitic worms and flukes, which are too large for phagocytes to engulf, by releasing an enzyme which can degrade fluke body walls. They are more effective if the worm or fluke has antibody attached.

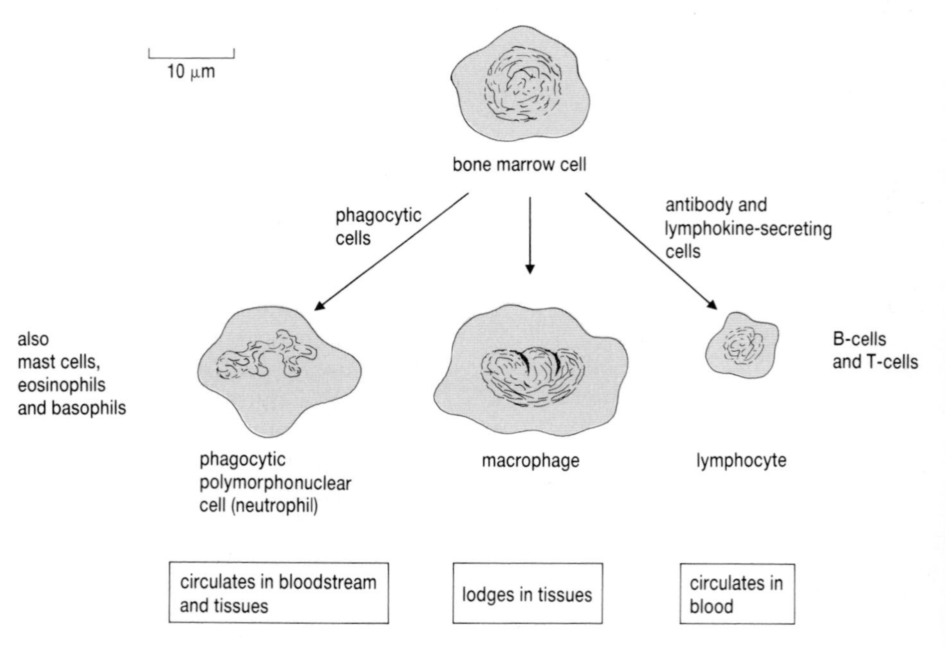

Fig 6.11 The structures of white cells.

Phagocytes

Phagocytes ingest and degrade particulate matter.

Phagocytes, shown in Fig 6.12, are scavenging cells found in the bloodstream and tissues which pick up living and dead organisms, cell debris and other particulate matter. They respond to a variety of signals found on micro-organisms and work better if the material they are scavenging is linked to antibody, complement or carries certain common bacterial components. There are two different sorts of phagocytes in most animals. **Polymorphonuclear leucocytes**, or neutrophils, are the commonest white cells in the blood. They circulate with the blood and can move into tissues. They are attracted to the site of damage or infection by a variety of chemicals, particularly histamine and small peptides released by

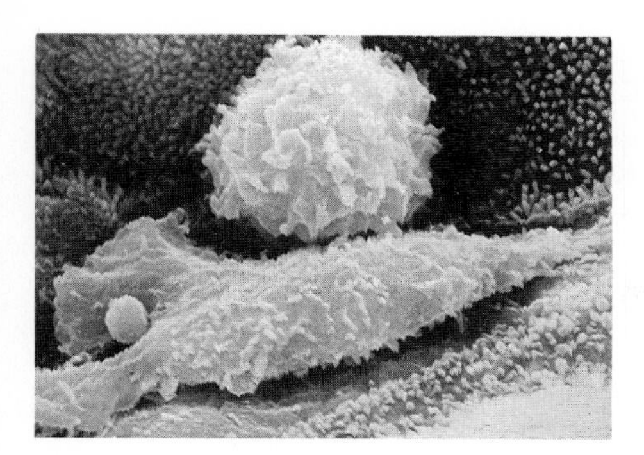

Fig 6.12 The nearer of these two macrophages is in the process of engulfing a particle. The 'arms' will eventually completely surround the particle enclosing it in a vacuole.

bacterial breakdown, then search for particulate matter to engulf. Receptors on phagocyte membranes bind with antibody or complement C3b attached to the material they are scavenging. This stimulates the phagocytic cell to engulf the particle by enclosing it in a vacuole, a process known as endocytosis. Lysosomes fuse with the vacuole, emptying in hydrolysing enzymes and enzymes which make toxic chemicals such as peroxides. These kill and degrade the vacuole contents.

Most bacteria are killed within half an hour but some organisms such as *Brucella* and *Mycobacterium tuberculosis* can endure these conditions and grow within the cells, causing chronic infections that are difficult to clear up. Some microorganisms are more resistant to being engulfed because they have a capsule or other chemicals on their surfaces that inhibit phagocytes. Phagocytes are very short-lived but are important in the first response to wounds and infections.

Macrophages are also phagocytic cells but are lodged permanently in tissues (shown in Fig 6.12). They develop from white cells in the blood known as monocytes but they quickly leave the circulatory system and settle in the tissues in areas where organisms could gain entry. They are common around the gut and in lung tissue, and also in the liver and spleen. Those in the liver and spleen and some other areas are known collectively as the **reticulo-endothelial system**. Macrophages are better able to ingest bacteria if they are triggered by lymphokines, which are chemicals released by a kind of lymphocyte called a T-cell. Macrophages last longer than phagocytes and are important in combating deep-seated or chronic infections.

Lymphocytes

Lymphocytes circulate in the blood and the lymphatic system. They are also sited in lymph nodes, the spleen and in other places interspersed with macrophages. Lymphocytes migrate from the blood to infection sites drawn by chemicals released by other cells. Lymphocytes have different functions though they all have the same structure. All lymphocytes start out as stem cells in the bone marrow, but they migrate into the bloodstream to finish their development elsewhere. **B-lymphocytes** are matured by various lymph nodes round the body and become antibody-secreting cells. When a B-cell is sensitised to an antigen it is triggered by a T-helper cell to multiply rapidly and make a clone of cells sensitised to that antigen. The cells circulate secreting large amounts of antibody to that antigen. Some of them, however, remain settled in the lymph node as memory-carrying cells. They may live for many years, only activated when re-exposed to the antigen.

Most lymphocytes are **T-cells**. These are produced by the bone marrow but develop under the influence of the thymus gland in the chest. The roles of the various types of T-cells are still not fully understood but are summarised in Table 6.4. Some T-cells, called T-helper cells and T-suppressor cells, regulate B-cells and T-cytotoxic cells, either by stimulating them into activity or suppressing them. Like the B-cells, T-helper cells are dedicated to one particular antigen. Sensitising T-helper cells to an antigen is a complicated process involving the main histocompatibility proteins found on each individual's cells.

T-cytotoxic cells kill infected cells which contain antigens such as viral proteins, and play a vital role in coping with infections. Antibodies in the blood hinder the spread of infection; T-cytotoxic cells destroy cells which are already infected. They detect virus proteins alongside the histocompatibility proteins in the infected cell membrane and use an enzyme to kill the cell. T-cytotoxic cells have a role in a healthy body too. They destroy faulty cells such as incipient tumour cells; they will also destroy transplanted tissues unless they are suppressed. A fourth kind of T-cell secretes lymphokines and chemicals that activate the complement system.

Table 6.4 Roles of lymphocytes

Cell	Function
B-cells	multiply to make plasma cells which secrete antibodies
memory cells	retain memory after initial encounter with antigen
T-helper cells	bind to antigen on macrophages, cause release of lymphokines which activate B- and T-cytotoxic cells and cause multiplication
T-suppressor cells	suppress B- and T-cytotoxic cells
T-cytotoxic cells	destroy virus-infected cells and cancer cells
T-cytotoxic memory cells	retain memory of antigens

QUESTIONS

6.18 Give one functional difference between polymorphonuclear phagocytes and macrophages.

6.19 Why do micro-organisms die when they are engulfed by a phagocytic cell?

6.20 Give one reason why a disease like TB is so difficult for the victim to deal with.

6.21 In an investigation of a particular species of pathogenic bacteria it was observed that the normal strain, which was very virulent, had a capsule but that a mutant strain, which was unable to make a capsule, could not infect healthy animals. Suggest two explanations for this observation.

6.22 Make a list of all the types of white cell mentioned in sections 6.5 and 6.6. Give one function for each of them.

6.7 PLANT DEFENCE MECHANISMS

Plants have evolved many different strategies to cope with the depredations of pathogenic organisms. Some plants contain pre-existing compounds which affect the damage-causing organisms. Other chemicals may be synthesised after an infection has started which inhibit pathogen activity. Even growth patterns may help a plant defend itself against infection. For example, the fungus causing ergot in some cereals such as barley is prevented from infecting wheat by self-pollination. The fungal spores must land on a stigma before pollination but wheat flowers do not open until after fertilisation.

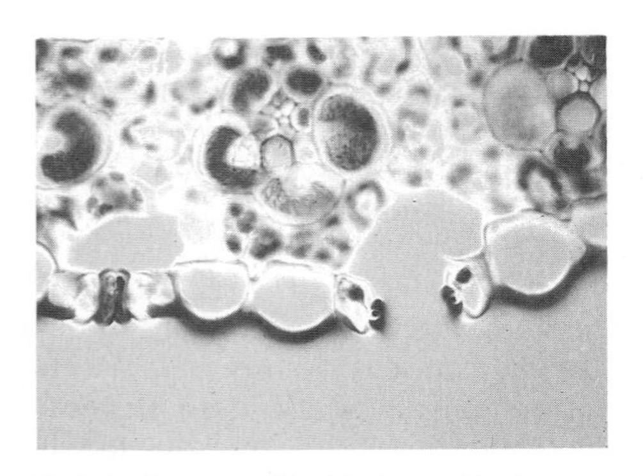

Fig 6.13 The waxy cuticle of the lower epidermis, stained to show as a thick layer, is a barrier to microbial entry. Open stomata, one is shown here, offer an easier route into the spaces of the leaf interior.

Natural barriers

Plants have physical barriers which make it hard for pathogens to get in. There are organisms which live on the surface of plants, called the phyllosphere flora, and pathogens have to compete with these if they are to gain a foothold. Once inside, however, there is far less competition. Fungal spores on leaves have to compete for scarce nutrients leached from the leaf for their growth after germination. Leaves are covered in a waxy cuticle, which can be seen in Fig 6.13, which is an effective barrier to many organisms. The cuticle may be thick or contain toxic chemicals. An easier way in is through one of the many natural openings in a plant's surface such as lenticels and stomata.

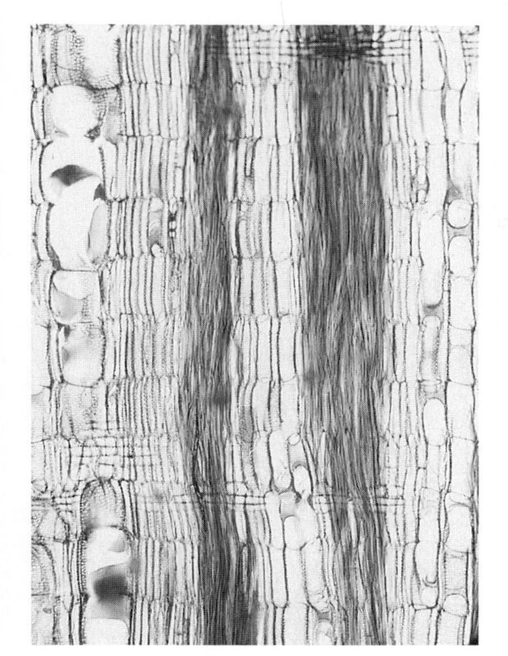

Fig 6.14 Tyloses, seen above, are cytoplasmic blockages in transporting vessels which bar the further movement of pathogens through a plant's transport system.

Non-specific chemicals

There are a variety of responses available to a plant if a pathogen does invade. Its spread can be blocked by substances which make a physical barrier around it. Viruses are spread passively, for example in fluid streams in phloem and xylem vessels. Fungi, however, tend to push the tips of their hyphae between cells as they grow. Lignin makes a very good barrier because it has great compressive strength and few organisms can degrade it, not even the plant producing it. Suberin is also a good barrier compound. Viruses as well as fungi are blocked if plants secrete gums blocking xylem and phloem vessels. Plants can also plug vessels with **tyloses** which are extrusions from neighbouring cells through the vessel wall. These can be seen in Fig 6.14. New vessels may differentiate to replace the lost ones.

Pathogenic bacteria secrete enzymes which degrade cell walls and cytoplasm. One response to this is to increase the calcium in pectin compounds in cell walls making them more resistant to the pectolytic enzymes. Another response, found in tulips, is that cells contain a compound which becomes toxic when acted on by fungal enzymes and inhibits the pathogen. Phenol-based compounds such as lignin precursors, catechol and saponins in cells may inactivate fungal enzymes or toxins. Tannins also inhibit virus multiplication. Resistant plants may have high levels of these compounds in their cells. An extreme response is **hypersensitivity**: the pathogen gets through the cuticle, but when it enters cells both the affected cells and those around it die. This blocks the development of the pathogen, though the mechanism is not fully understood. When the cells have died a layer of blocking compounds may be formed around them.

Phytoalexins

Phytoalexins are compounds with some anti-microbial activity which are formed by plants as a response to cell damage. Many pathogens exude compounds called **elicitors** as they grow which provoke phytoalexin production by the host plant. Phytoalexins are produced by cells near the site of infection, but they are not specific to one pathogen as antibodies are. Many plants have been shown to make phytoalexins, though they have not been found in some important crop plants; different phytoalexins are made by different species of plants. They are mainly anti-fungal and there are some anti-bacterial phytoalexins but they do not affect viruses. The commonest are chemicals called flavonoids and terpenoids. Table 6. 5 lists some phytoalexins found in common plants.

Phytoalexins are effective in very small amounts, 1/10 000 mole or less, but how they work remains a mystery. They can sometimes work by killing the plant's own cells, thus reducing the nutrients available to the pathogen and releasing toxic compounds. Some fungi can evade the effects of phytoalexins either by avoiding provoking phytoalexin production or by being able to degrade them.

Table 6.5 Phytoalexins from some common plants

Phytoalexin	Plant
wyerone	broad bean, *Vicia fabae*; lentil, *Lens*
medicarpin	broad bean; clovers, *Trifolium* spp
hemigossy pol	cotton, *Gossypium*
phaseollin	French bean, *Phaseolus*
orchinol	orchids
capsidiol	red pepper, *Capsicum*
lathodoratin	sweet pea, *Lathyrus*
pisitin	garden pea, *Pisum*

QUESTIONS

6.23 How does the surface of a leaf protect against infection?

6.24 What is a phytoalexin?

6.25 What is an elicitor?

6.26 Antibodies are said to be specific but phytoalexins are not. What does this mean?

6.27 Name two chemicals found in plant cells which may protect against plant pathogens.

6.28 Give an example of a compound which may be more abundant as a result of a pathogen infecting a plant.

6.8 PROVOKING BETTER DEFENCES: VACCINES AND ANTISERA

An infection raises an immune response in an animal which protects against further infection, however an immune response can also be raised artificially. Provoking an immune response to a disease is called **immunisation.** Artificial immunisation was first used in Britain in the eighteenth century when cowpox virus was used to stimulate the production of antibodies and lymphocytes which would respond to antigens on smallpox viruses. The process of protecting against smallpox was called **vaccination** after the Latin word for 'cow'. Now many other microorganisms are used to raise an immune response and reduce the incidence of disease. They are administered in the form of microbial suspensions called **vaccines**.

Active immunisation stimulates the body's own defence mechanisms. Passive immunisation is the donation of antibodies from another source.

Active immunisation occurs when a vaccine stimulates an individual to produce antibodies and a memory response dedicated to the organisms in the vaccine. Not all diseases can be prevented by vaccines however; some organisms simply don't provoke a strong immune response. Nevertheless antibodies may still be used to treat or prevent an infection. **Passive immunisation** is the injection of a solution of antibodies, or **antiserum**, into an individual to protect against an infection for a short time. Antisera give immediate protection whereas vaccines take several days to be effective.

Fig 6.15 Vaccination with a jet injector, using compressed air to introduce vaccine into the skin layers.

Vaccines

There are over 50 human vaccines in common use in the developed countries giving some protection against a variety of bacterial, viral, protozoal and helminth diseases. As yet there is no similar preventative that can be used for plants. Vaccines are made from fragments of organisms, modified toxins, organisms killed by treatment with a variety of chemicals or from live, weakened strains of the organisms; all of these carry antigens. Weakened strains are described as being **attenuated**, that is they provoke the immune system but do not cause as severe an infection as the normal strains. Vaccination, shown in Fig 6.15, may be by injection into the bloodstream or inoculation into the skin layers or muscle, or be administered in a drop of liquid as an oral vaccine. Oral vaccines have components which can withstand passage through the stomach and be absorbed through the gut wall.

The effectiveness of a vaccine depends on a number of factors. Some antigens are better at provoking a response than others; weak antigens may be mixed with adjuvants to provoke a stronger response. The number of different strains in a species of pathogen is important; a vaccine needs to carry antigens from all the common strains to be useful. Not all vaccines are trouble-free – some cause side effects because of cell components present in the vaccine, but the new vaccines made using engineered bacteria should reduce the problems. There is also a slight risk using attenuated strains: a tiny minority of the population, about one in three hundred thousand recipients, may suffer severe infections, but this risk is very low compared to the risks involved in suffering the infections. The vaccine manufacturing process is covered in Chapter 5.

Antisera

An antiserum is a solution of antibodies which is administered to reduce the effects of an infection or a microbial toxin. It lasts for about three weeks and then declines quickly, so it is a short-term measure. Antiserum is made by injecting a suitable animal with an antigen and provoking the production of antibody to it. The plasma containing antibodies is concentrated and used to protect people who have been exposed to a pathogen carrying the antigen. It is useful in treating infections like hepatitis B for which there are no really effective drugs. Some people develop reactions to animal antiserum, so it cannot be used for everyone.

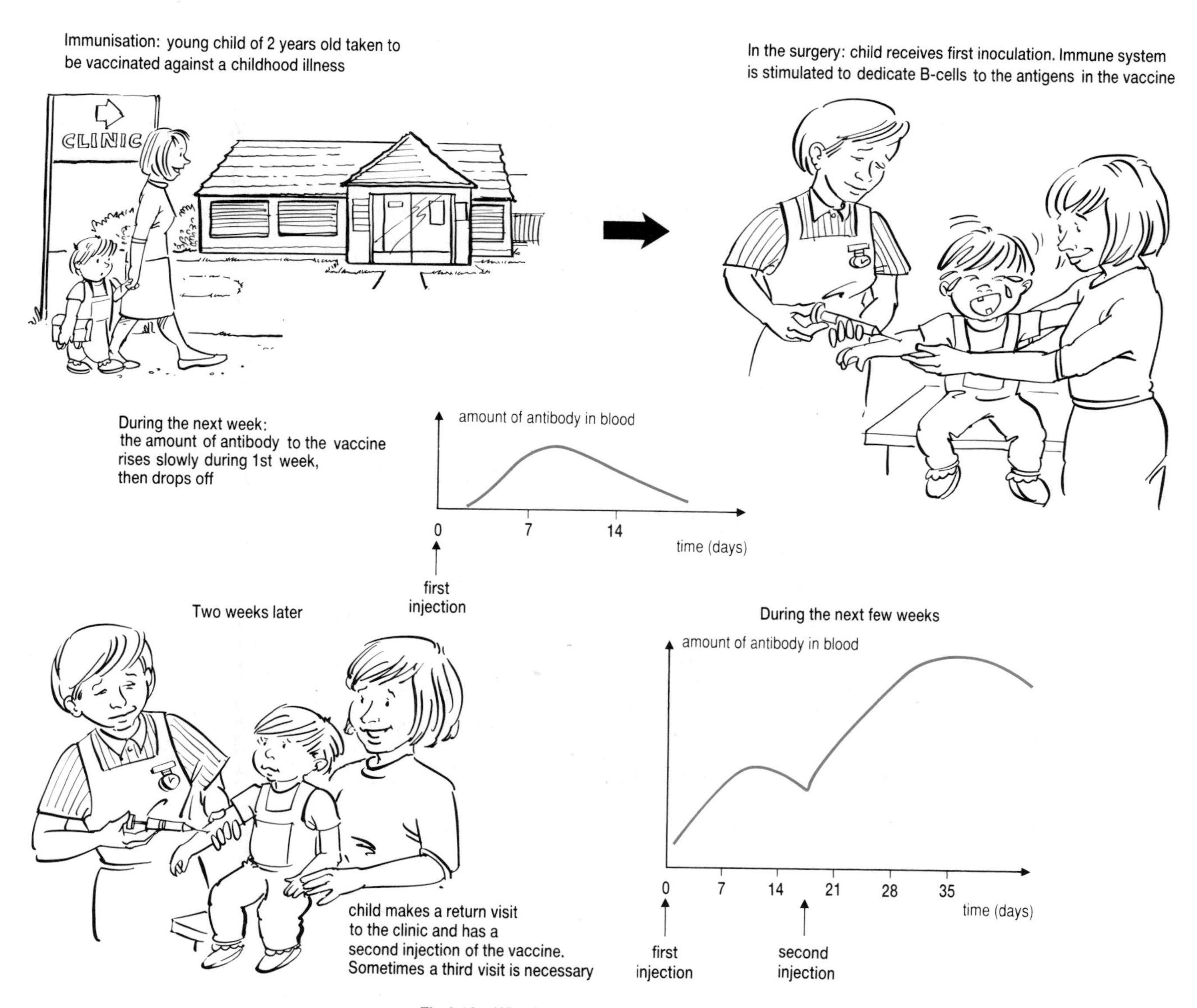

Fig 6.16 What happens when you are vaccinated.

Human antibodies can be collected from the blood of people who have recovered from a disease, or from a general pooled human serum made from the combined plasma. For many years this was the only way in which Lassa fever could be treated. Now specific monoclonal antibodies can be made and used to treat particular diseases.

Immunisation

If a person has not met a particular micro-organism they will not have antibody or lymphocytes to it, but they can be immunised. The procedure is shown in Fig 6.16. After the first inoculation the primary response develops. Specific antibody, which is mainly IgM, starts to appear in the blood in increasing amounts, or titre, over the next few days. It is followed up with a second inoculation between four and six weeks later to get a secondary response. The antibody titre is quite low after the first injection but rises to very high levels within 48 hours of the second injection and lasts a long time. After the second injection large amounts of IgG are produced as well as IgM. Some vaccines, for example the anti-tetanus vaccine, may need a booster dose from time to time to maintain the secondary response.

Vaccination policies

Most countries have a policy of vaccinating young children against the childhood diseases that are still are major killers. The programme usually includes diphtheria, whooping cough, polio, tetanus, and measles. Even so, 3.75 million children worldwide (not including China) die each year from measles, tetanus and whooping cough. The organisms causing TB and rubella are also vaccinated against because TB takes a long time to treat and because a rubella infection of a pregnant woman can lead to extensive damage to the unborn child's nervous system.

The inoculations are given at different stages of a child's life starting when the child is four to six months old. When a child is very young it is protected by maternal antibodies passed across the placenta or via the breast just after birth. If these antibodies are still in circulation the child will not make an effective response to the vaccine. The strategy reduces the numbers of children catching a disease and the numbers suffering severe complications.

Not all children are vaccinated, even in countries with well-developed primary health care services. Some children are not vaccinated because they may suffer side effects. Others do not finish their course of vaccinations for a number of reasons. Some children in developed countries are not vaccinated because their parents are unaware that their children could suffer; in developed countries people seldom see severe cases of childhood infectious diseases and assume that the infections have died out or are not likely to infect their own children. Many are unaware of the complications that can occur with what seem to be quite minor diseases.

Table 6.6 Notifications and mortality of measles in England & Wales. From OPCS, Communicable Disease Statistics 1986

Year	Cases	Deaths
1967	460 407	100
1968	236 154	51
1970	307 408	42
1972	145 916	28
1974	109 636	20
1976	55 502	14
1978	124 067	20
1980	139 487	26
1982	94 200	13
1984	62 080	10
1986	82 078	–

Worldwide, over two million children die and many more are handicapped each year by measles, a preventable disease. Even now there are regular measles epidemics in Britain and children do die. The number of cases of measles, and the mortality rate, in England and Wales since 1967 is shown in Table 6.6. The use of measles vaccine has become more common during that time; in the USA all children must be vaccinated against measles before starting school.

Not all of a population needs to be vaccinated in order to break the transmission chain.

Even if only a proportion of children are vaccinated, diseases can be controlled if the number of potential victims is low. As long as at least 60 per cent of the population is vaccinated against a disease it will remain as a sporadic infection. If the level of community protection falls below that then more serious epidemics can occur.

ENCOURAGING VACCINATION

Fig 6.17 A group of women in Nigeria learning about smallpox and similar diseases as part of a vaccination and health improvement campaign.

The National Health Service in Britain was inaugurated in 1948. Before that time people had to pay for any medical treatment, so the poorer sections of the community were reluctant to visit doctors. Old habits were hard to break and the campaigns to get children vaccinated had to overcome this reluctance. Many strategies were used to encourage parents to get their children vaccinated. In 1952 the health authorities in Birmingham began a regular programme to encourage vaccination. At that time, many foods were still rationed and people needed a ration card in order to get them. By giving everyone applying for an egg ration for six- month-old children a leaflet about vaccination, contact with parents was ensured. Another leaflet and consent card were sent a month later, and birthday cards from the Central Council for Health Education were sent on the first birthday of every child born in the city. Health visitors checked children, and a reminder card was sent for any child who had not started on a vaccination course by the time it was two-years-old. Another reminder card went to those who had only got as far as primary immunisation just before they were five-years-old and due to start school. All the leaflets and cards stressed that the visits to the clinics cost nothing.

QUESTIONS

6.29 Distinguish between active and passive immunisation.

6.30 What is a vaccine? Review section 5.5 and summarise vaccine manufacture.

6.31 Examine Fig 6.18 which shows the events which occur in the human body over a period of time following the administration of a vaccine such as the typhoid vaccine. Explain the body's response to the administration of the vaccine
(a) in the region indicated by A,
(b) after the secondary immunisation.

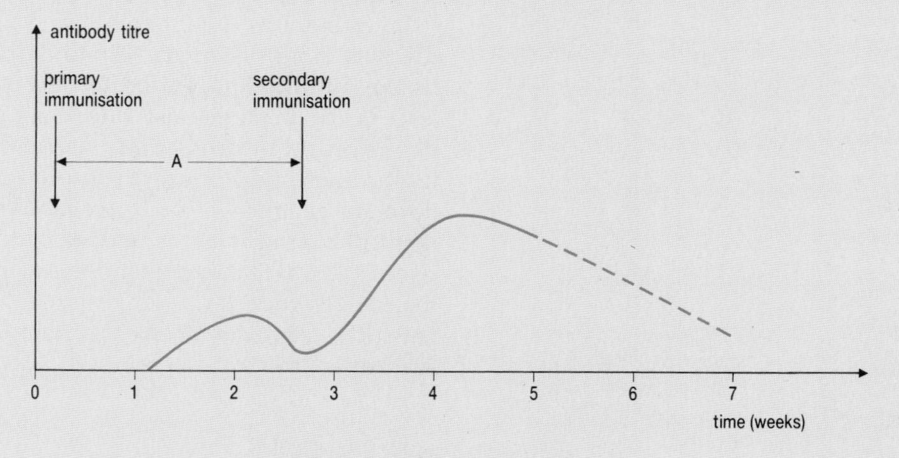

Fig 6.18 Change in antibody titre.

6.32 Examine Table 6.6, then use the figures to construct a graph of the incidence of measles over the period 1967-86.
(a) Comment on the figures.
(b) Is there a correlation between the number of cases and the number of deaths?
(c) Is the chance of a child dying the only cause for concern in a measles infection?

Fig 6.19 For many years the cinchona tree was known as Jesuits' Bark, because Jesuit missionaries were the first Europeans to observe the use of the quinine-containing bark extract in South America. The missionaries took it with them as they travelled round the world, introducing the tree or the medicine made from it.

Chemotherapy is the use of drugs to treat disease. There is a long history of the use of plant extracts to treat illnesses but until recently the results were inconsistent, and no-one understood the basis on which they worked. Many plant extracts have now been characterised and better understanding of the active ingredient and how it works has improved their use. For example the bark of the cinchona tree, which can be seen in Fig 6.19, was used over three hundred years ago to treat malaria, a protozoal disease; now the active ingredient quinine is extracted and other therapeutic compounds containing it are made.

There has been a parallel development in the use of chemicals to control plant pathogens, but a sick plant is generally abandoned to its fate or destroyed; most effort is devoted to preventing other plants from acquiring the infection.

Drugs

At the turn of this century there were no effective cures for common human infections, neither was disinfection widespread though it had been discovered, so many patients died of diseases and septic infections. The earliest development of synthetic therapeutic agents came with the work of Paul Ehrlich who was looking for chemicals that could kill pathogens. In 1909 he and Hata worked on an organo-arsenic compound Ehrlich had synthesised which was effective against syphilis bacteria. It was used as the first synthetic chemotherapeutic agent for people, though it did have side effects. Once a chemical had been found that selectively killed bacteria others were sought. Within a few years the first sulphonamide drug was developed and many more have been discovered since.

Penicillin, an antibiotic, was discovered shortly after the first synthetic drugs were developed, though it was not used therapeutically for 15 years after the first observations. A whole range of drugs have been developed since then. Anti-viral agents were not discovered until much later and there are still few effective anti-viral drugs available.

Investigations were almost random at first but now the search for new drugs is more systematic. An understanding of cell biochemistry has given us the information needed to find chemicals which will interfere with the activity of perhaps just one enzyme. A promising new line is in the development of drugs that prevent the proper coiling of bacterial DNA in its chromosome. As human chromosomes are coiled in a different way with different chemicals the drugs should not affect human cells. The development of a drug is an expensive and time-consuming business; it may cost over $100 000 000 to develop a new drug and take 10 years before it can be marketed. Very few compounds that show anti-microbial activity ever reach the chemists' shelves as many are too expensive to make on a large scale or are too toxic to use.

Drug names

Every drug has a systematic chemical name which is often long and complicated. Each manufacturer of a particular drug also uses a commercial name for it, and manufacturers may sell the same drug under different names in different parts of the world. The commercial name that a drug is sold under is a trademark which is the property of the manufacturer and cannot be used by other people. To make life easier for people dispensing or prescribing drugs many have an international nonproprietary name (INN) that can be used by everyone and is much shorter. The INN includes the kind of drug it is and its type of action. For example all the penicillin derivatives are names ending in -cillin.

Antibiotics

An **antibiotic** is a compound produced by a micro-organism which inhibits the growth of other micro-organisms. Antibiotic activity was first observed scientifically in 1929, when Alexander Fleming observed that a fungus, *Penicillium notatum*, inhibited the growth of bacteria on an agar plate. It was many years before the compound responsible for the inhibition was isolated and identified, then even longer before reasonable quantities of penicillin were obtained. Trials of penicillin in humans took place in the early 1940s and it was brought into general use after 1945. Thousands of antibiotics have now been discovered, but relatively few are used as medicines. Many are too toxic to use, or they are not as effective as compounds already in use, or they only affect organisms that do not cause great problems and are therefore uneconomical to produce. There are also some potentially useful antibiotics that cannot be made successfully on a large scale.

Table 6.7 Chemotherapeutic agents

Name/chemical type	Activity/use
erythromycin	antibiotic, inhibits 50S, ribosome sub-unit
penicillin, a β–lactam	antibiotic, blocks cells wall synthesis in some bacteria
nystatin	antibiotic with anti-fungal effect in animals from *Streptomyces noursei*
sulphonamides sulphamethoxazole	analogue to 4-aminobenzoic acid, a coenzyme, e.g. needed in bacterial metabolism
trimethoprim	interferes with bacterial folic acid sythesis
tetracycline	antibiotic from *Streptomyces aureofaciens*, prevents tRNA linking to ribosomes
amphotericin B	anti-yeast activity in animals
pentamidine	anti-protozoal activity, treats sleeping sickness
tryparsamide	arsenic compound used in late stages of sleeping sickness
streptomycin	antibiotic from *S. griseus* with anti-TB activity, inhibits correct amino acid incorporation in protein molecules
chloramphenicol	antibiotic which blocks peptide bond synthesis
para-aminosalicylic acid	anti-TB activity
rifampicin	antibiotic, inhibits mRNA synthesis

There are many different sorts of antibiotics. Most work by inhibiting a biochemical pathway in the target micro-organism, blocking its growth. Some aspects of metabolism are unique to micro-organisms so an inhibiting antibiotic would not have adverse effects on human or animal cells directly. Other antibiotics may affect processes common to animals and micro-organisms; these have to be used much more carefully. Antibiotics can cause adverse reactions in people for a variety of reasons, for example a significant proportion of the population is allergic to penicillin and its derivatives. Most antibiotics work on bacteria; there are few anti-fungal or anti-viral compounds.

Antibiotics are not equally effective against all bacteria. An antibiotic will affect only those bacteria which have a particular property in common. For instance, penicillin, a β-lactam, inhibits the synthesis of peptidoglycans in new bacterial cell walls resulting in a loss of strength. The bacterium eventually suffers osmotic lysis. Only the Gram positive bacteria, which have a significant amount of peptidoglycan, will be affected by penicillin; the remainder will be unaffected. An antibiotic that is effective against a wide range of organisms is called a **broad spectrum** antibiotic. There are many of these, for example tetracyclines and chloramphenicol. Antibiotics which affect only a few species are described as **narrow spectrum**.

Treatments for plants

The chemicals used against plant pathogens are used to control the spread rather than to cure individual plants. There are no effective anti-viral compounds that can be used on a large scale. Traditional agricultural practices such as dusting with sulphur, calomel and Bordeaux mixture, which is a copper compound, were in use as fungicides while Pasteur was still working on his germ theory of disease. Most are used as surface-active agents which kill on contact, that is they lie on the surface of the plant as a dust or liquid spray and kill fungal spores or hyphae growing over the surface, (see Fig 6.20). Organo-mercury fungicides were developed at the beginning of this century and organo-sulphur compounds, the **dithiocarbamates**, in the 1930s. These could be used on a wider range of plants and had fewer side effects. They are used as preventative measures, for spraying crops when pathogen infection is likely, or as seed and root dressings before planting. Some chemicals are absorbed by the plant and then translocated. These **systemic** agents protect against pathogens wherever they infect and include compounds such as **benomyl** and thiophenate.

Fig 6.20 Surface-acting fungicides are sprayed onto the surface of the plant to inhibit any air- or water-borne spores which may land on the foliage. Early warning systems allow farmers to judge when there is threat of infection and to apply fungicide only when it is needed.

Fungicides may be wide spectrum, that is they kill many different species of fungi, or narrow spectrum, only killing a few species, for example tridemorph kills only cereal mildews. Fungicidal powders are mixed with other materials before they are used: they are made wettable and are mixed with spreaders which enable the fungicide to be sprayed as an even coating that sticks onto the plants. As fungicides are cheap and widely available little work has been done on other methods of control.

Drug resistance

Many pathogens become resistant to drugs by, for example developing enzymes that degrade a drug or inactivate it. Drugs usually work by inhibiting a particular step in a biochemical pathway in the micro-organism. If an alternative pathway is available, or is evolved, the drug will no longer work. Penicillin resistance is due to an enzyme, β-lactamase, which degrades penicillin. The genes for β-lactamase are housed on a plasmid that is carried by resistant bacteria. Unfortunately, treating infections with drugs selects for those organisms that can resist them. These survive treatment, even though 99.9 per cent of their fellow invaders may be killed or inhibited. The survivors then continue to multiply. Many of these are destroyed by body defences but some may escape and continue to cause infections. If they are transmitted to another victim they may spread further and very quickly they will dominate the local population. Drug-resistance plasmids can often be transferred between bacteria of the same or different species. This phenomenon is known as **transferable drug resistance**.

The more widely a drug is used the more likely it is that resistance will develop. For instance, the widespread use of aminoglycoside antibiotics (streptamycin, etc) by doctors and hospitals to treat infections of *Staphylococcus pyogenes aureus* and *Pseudomonas aeruginosa*, which cause sepsis, has led to resistance. In Austria, by 1985 it was found 42 per cent of strains of *S. pyogenes aureus* were resistant to the common aminoglycosides such as gentamicin. In contrast, following an outcry in Scandinavia when it was realised what was happening, the use of these drugs was restricted and resistance has settled at 2–3 per cent.

Resistant varieties of many major pathogens are now common. The main way to deal with the problem is to use a treatment of two or three drugs combined together, each acting in a different way. Though a strain of pathogen may develop resistance to one drug it is extremely unlikely that it will be able to evolve resistance to three different drugs acting in three different ways simultaneously. For example, leprosy has been treated with dapsone for over 40 years, but now there are resistant strains in many parts of the world. The newer treatment is to use dapsone together with two other drugs, rifampicin and clofazimine as a triple drug.

At one time antibiotics were given to farm animals routinely as a **prophylactic** to prevent infections but it was soon realised that this encourages drug resistance. When a farm animal, such as a cow, is given an antibiotic in its feed the sensitive bacteria will die but resistant ones will be unaffected. The resistant bacteria will thrive in the reduced competition and reach large numbers. Plasmid-carried resistance can then be transferred to other bacteria, particularly harmless species, and spread. Before long the resistance is acquired by a number of harmful organisms so that when the cow does succumb to an infection the antibiotic is no longer effective. As a result of the increase in antibiotic resistance, the practice has stopped and antibiotics are now only used therapeutically.

QUESTIONS

6.33 Distinguish between a 'chemotherapeutic agent', a 'prophylactic' and an 'antibiotic'.

6.34 What is the difference between a broad spectrum and a narrow spectrum antibiotic?

6.35 What is transferable drug resistance and why is it a problem?

6.36 Name an anti-fungal compound used for plant infections.

6.37 What is the usual treatment for an infected plant?

6.38 Alexander Fleming first observed the effect of an antibiotic when he noticed that bacteria did not grow around the site of penicillin production. How could you use this observation to compare the effect of the antibiotic streptomycin on the growth of two different bacteria?

SUMMARY

All groups of micro-organisms have pathogenic members. Bacteria and viruses are most important in animal disease; fungi and viruses in plant disease. Pathogens may cause cell and tissue damage, release toxins or affect host cell activity, reducing the ability of the host to survive.

Micro-organisms can be transmitted by wind, water, soil or food, or inoculated directly into blood or sap. Some micro-organisms exploit vectors to gain access to new hosts.

Animals and plants have a range of natural barriers which hinder the entry of pathogens and can induce chemicals to inhibit pathogen activity. Animals, but not plants, can induce a response specific to a species of pathogen which provides the animal with immunity. A range of chemicals, some natural some synthetic, have been developed to treat microbial infections.

Chapter 7

PLANT DISEASE

LEARNING OBJECTIVES

After studying this chapter you should be able to:

1. describe the main causes of plant disease;
2. understand a range of plant diseases;
3. give a detailed description of selected plant diseases;
4. describe the control methods used against plant disease;
5. explain how disease-free stock plants can be produced.

7.1 INTRODUCTION

Plant diseases were recognised as being infectious at the time that microscopes were developed. Rose rust was one of the earliest specimens to be examined by the new invention; Fig 7.1 shows Robert Hooke's drawing published in 1665. Fungal spores were identified and the role of fungi in decay was recognised. Mercury compounds were used as wood preservatives at that time, but it was not until the nineteenth century that fungicides came into use. Forsyth, who was King George IV's gardener, wrote of using an infusion of sulphur, lime and elder buds to make a preventative against mildew on fruit trees. This became known as lime-sulphur and was used on a variety of diseases afflicting plants as different as onions and grape vines. The outbreak of potato blight in Europe in the nineteenth century focussed attention on plant disease and the link between the blight and a fungus was finally established by de Bary in 1863.

Fig 7.1 Robert Hooke's drawing of rose rust is one of the earliest illustrations of a pathogen. The bitrophic fungus, *Phragmidium* has spores in a gelatinous sheath on the end of a long stalk which can be seen clearly in the drawing.

Plant diseases are as important as animal diseases. It is difficult to assess how much damage is done by plant pathogens but about $5 billion are spent each year on trying to prevent plant disease. It has been estimated that if plant disease treatments were not used agricultural output could decline by as much as a third in many countries. Epidemics of diseases such as potato blight and coffee rust, both fungal infections, have had major consequences for the economies of whole countries. The Irish potato famine of the 1840s led to the death of a million people and the emigration of many more, while coffee rust epidemics in India and Sri Lanka last century wiped out entire coffee plantations and, it is said, turned the British into tea drinkers. Some infections are confined to a particular country but many more are endemic in many parts of the world.

7.2 INDUCTIVE AGENTS

Plant disease is caused by as wide a range of factors as animal disease. As well as pathogenic micro-organisms, disease is caused by nutrient deficiencies, physiological malfunctions and environmental factors such as frost. The most important pathogens are fungi: over 1000 fungal species have been found to be associated with plant disease. Viruses and mycoplasmas are also important pathogens and there are some plant diseases which are caused by agents smaller than viruses known as satellite viruses and viroids. Bacteria cause few problems unless they infect after damage from another cause. Bacteria are poor plant invaders because they cannot push through the leaf and stem cuticle or through plant tissues, and they cannot move far independently. Also the acidic interior of a plant does not suit most bacteria. A few diseases, for example the red rust that parasitises citrus plants, are caused by algae.

Fungi and viruses are the most important plant pathogens.

Many pathogens are entirely dependent on a living host plant, and so debilitate rather than kill their host. Some fungal pathogens are not completely dependent on a host. They are opportunist, living in the soil saprophytically on dead vegetation but taking advantage of plants weakened by damage or another infection. *Penicillium* and *Rhizopus*, for example, are both fungi which cannot infect healthy fruit but will grow in wounded apples and cause fruit rot. In contrast, *Verticillium* and *Fusarium* species are obligate pathogens which need living plants for their life cycles, but eventually kill them. They then feed on the remains before infecting another plant.

7.3 DISEASE SPREAD

The source of infection of a crop is usually a wild plant infected with the pathogen, or else other crop plants with the infection. Often the pathogen is quite innocuous in the wild plant but causes major diseases in the crop. Pathogens are transmitted to crops when wild plants grow among them or at the edges of fields. Fungi spread by air- and water-dispersed spores but other pathogens need to be carried to a susceptible host by a vector or even by the grower on tools, wheels and boots. Fungal spores may be carried hundreds of kilometres in air currents even across oceans. Coffee rust, *Hemeleia vastatrix*, spread this way from Sri Lanka to Africa then on to Brazil where it caused enormous problems. If the weather conditions are favourable, that is warm and wet, the fungi may grow and produce spores in a very short time resulting in rapid spread of the infection.

A population of wild plants is very diverse with a range of characteristics affecting susceptibility to disease. It is usual for most plants in a population to be resistant to a particular pathogen as they have characteristics which enable them to overcome pathogen activity. These features, such as a thick cuticle or toxic compounds in cells, are carried genetically. However in cultivated plants there is less genetic diversity. Selective breeding programmes to improve yield or weather resistance often lead to the loss of genes for disease-resistance characters, so these commercial varieties are much more susceptible to diseases than their wild counterparts.

Fig 7.2 One cultivar of grain covers 23 acres of farmland in Oxfordshire. The grain harvest that year broke records; but any plant pathogen which could initiate an infection in this monoculture could have devasting effects, it would be able to infect every plant in the picture, except the trees.

Farming practices such as **monoculture**, which is the growth of a single strain or **cultivar** of a crop plant over a large area, make the situation worse. Fig 7.2 shows an area turned over to the cultivation of a single variety of grain. Monoculture is used because genetically identical seed enables the farmer or horticulturist to get the consistent quality demanded by consumers of the crop. Unfortunately any pathogen that can infect spreads rapidly through acres of identical plants causing enormous damage and loss of crop. The term 'boom and bust' was coined to describe what happens: a resistant cultivar of plant is developed and is seized upon by farmers. As it needs less pesticide treatment it is cheaper to grow and yields are higher, so for a few years the crop booms. Then the inevitable happens, and a pathogen evolves a method of overcoming the plant resistance and wipes out whole crops – the bubble bursts.

Plant varieties are now grown in places where they do not occur in the wild and may meet new pathogens which they have no resistance to. Similarly when seeds, tubers or fruit are exported from one country to another a variety of animal and microbial life may go with them. These unsuspected migrants can cause havoc among the plant populations of their new host country. Dutch elm disease, discussed in section 7.5, is an example of a plant pest that arrived in a new country where the native plants had no resistance and spread rapidly through the population.

QUESTIONS

7.1 Give one cause of plant disease other than micro-organisms.

7.2 What is meant by the term 'monoculture'?

7.3 Why does monoculture encourage the spread of disease?

7.4 Give three sources of infection of crop plants.

7.4 HOW PATHOGENS CAUSE DAMAGE

Not all pathogens are equally damaging. Some highly specialised fungi cause little damage apart from draining the plant's nutrient stores, whereas others are quite catastrophic causing a widespread derangement of plant activity. Though roots are exposed to spores and bacteria in the soil, leaves are the most likely parts to be infected. Leaves have a very large surface area exposed to air and water splashes which could carry fungal spores, and are refuges or food sources for a host of animal life. They are protected by a waxy cuticle with compounds such as gallic acid and terpenes which inhibit the growth of pathogens, but many pathogens have developed methods of penetrating this barrier. Some degrade the waxes, others use infection pegs or apressoria to penetrate the cuticle. Fig 7.3 shows an infection peg pushing through a leaf surface.

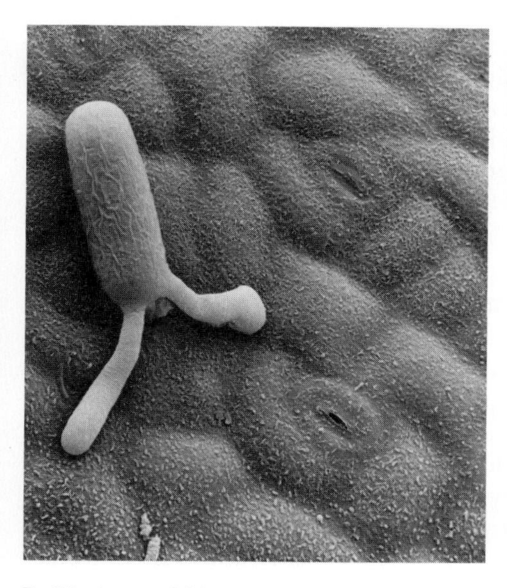

Fig 7.3 A spore of *Erisyphe pisi* initiating an infection in a pea leaf. The hypha on the right has a slight swelling, or appressorium, at the end; this will push through the leaf surface. A second hypha has begun to grow out. Some pathogens push through stoma (seen right) to enter leaves.

Tissue damage

Inside a plant most tissues are vulnerable to infections. Infections in transporting tissues, called wilt diseases, affect the movement of minerals, water and organic material. They also damage other tissues which are not infected but which are deprived of water and solutes, and so the whole plant droops and wilts like that in Fig 7.4. Some pathogens secrete enzymes such as pectinases to break down the plant tissues producing soft rots, and shoots die back. Bacterial pathogens cause damage to cell membranes and the tissues become waterlogged which makes bacterial movement much easier. Once the outer protective layers of the plant are breached organisms which may not be able to infect an intact plant can enter. Bacteria may set up secondary infections and fungi may invade from the surrounding soil.

Fig 7.4 The centre tomato plant is infected with *Verticillium albo-atrum*, a wilt disease; the others are healthy with turgid leaves.

Toxins

Plant pathogens may kill cells, produce toxic chemicals or disturb normal cell activity and nutrient storage.

Toxins are not as widespread as in animal disease, but there are a few. They may affect cells nearby, killing them for later colonisation, or be carried through the transpiration stream. *Alternaria*, a fungus which causes a pear black spot, and *Helminthosporium victotiae* which causes a South American oat blight both make toxins that cause the symptoms of the infection. Some vascular wilt diseases make toxins, for example *Fusarium* infections produce a compound, fusaric acid, that produces some of the wilt symptoms. Vascular wilts are usually caused by fungi but there are also some bacteria which infect transporting tissues. However, the substances made by bacteria usually block the transporting vessels so these are not true toxins.

The effects of plant disease

The physiological drain on a plant is often the most serious consequence of plant disease from our point of view. Virus activity reduces chlorophyll production, hence photosynthesis and the yield of a crop. A pathogen which kills potato leaves will reduce the number or size of potatoes made by a plant and the yield goes down. Many fungi divert the normal translocation of carbohydrates away from growing points, seeds and roots for their own needs or block transporting tissues. Photosynthesised materials will either stay in the infected leaf or be drawn into the infected leaf from elsewhere for fungal growth.

Fig 7.5 Peach leaf curl is caused by a fungus, *Taphrina deformans*, which grows between the cells in the mesophyll layer. It causes extra cell divisions and cell enlargement making the leak distorted and discoloured.

Some pathogens cause distorted growth so plants have leaves like those in Fig 7.5. The most obvious infection of this sort is *Agrobacterium tumifaciens* which causes crown galls. The pathogen provokes cell proliferation which makes a large, unsightly mass, known to generations of children as 'witches nests'. However there are other infections that cause distortion either through unusual cell division, such as peach leaf curl, or by interference with plant growth regulators.

QUESTIONS

7.5 Why is a leaf exposed to the air more likely to be infected than other parts of plants?

7.6 Give three symptoms that you might see which would indicate that your favourite houseplant had a fungal or viral infection.

7.7 Why could a pathogen growing in xylem vessels in the stem of a plant cause damage to a leaf?

7.5 FUNGAL DISEASES

Fungi are the most frequent cause of plant disease and sometimes cause huge crop losses. Many are distributed worldwide, for example *Puccinia striiformis* which causes the world's most common cereal rust. Fungal spores need a high humidity and a supply of nutrients to grow. A number of fungi infect through the root system of a plant or the stem close to the soil surface, but many fungal pathogens enter through the leaf surface. When a spore of a species entering through a leaf germinates, its hyphae spread across the surface of the leaf and absorb nutrients which have leached from the leaf. Some species have mechanisms to orientate themselves and allow them to recognise entry points. At the penetration points hyphae form a swelling called an **appressorium** (see Fig 7.3). This may be above a stomata or the hyphae may use pressure or enzymes to push an infection peg through the cuticle. Once inside a leaf the hyphae grow through and between cells.

Fungal pathogens include mildews, rusts, smuts, wound rots, soft rots and other diseases, Table 7.1 lists some important pathogens. There is one physiological grouping which reflects how good a parasite the pathogen is, that is the division into biotrophs and necrotrophs.

Table 7.1 Some fungal pathogens

Pathogen	Plants infected	Problem caused
Armellariella mellea	trees and shrubs	honey fungus, heart and root rot
Botrytis cinerea	many plants	grey mould destroys tissues, die back
Claviceps purpurea	cereals	ergot, grain infected
Diplocarpon rosae	roses	black spot, leaf damage
Erisyphe graminis	cereals	powdery mildew reduces photosynthesis and yield
Puccinia graminis	cereals	rust, reduces photosynthesis
Pythium	seedlings	damping off, seed rots in soil or stem attacked
Taphrina deformans	peach trees	leaf curl
Venturia inaequalis	apples	apple scab, blotched fruit spotted leaves

Biotrophs and necrotrophs

Fungal pathogens can be divided into those which kill their hosts quickly and those which drain them very slowly. Necrotrophs are unspecialised, opportunistic pathogens which infect a variety of plants, quickly disrupting the host's metabolism causing death. They infect damaged plants, weakened cells or ageing leaves, rather than healthy mature tissues, seldom spreading beyond the weak tissues. Where they colonise they produce degrading enzymes and toxins which cause the death of cells around the infection. Many necrotrophs can continue to live off the tissues after the plant has died and some may be able to grow on dead organic material if there is no other suitable host available. Examples include *Pythium* spp. and *Botrytis fabae*.

Table 7.2 Biotrophic and necrotrophic fungal diseases

Pathogen	Disease
Biotrophs	
Puccinia striiformis and *P. recondita*	cereal rust
Erysiphe	powdery mildews of cereals
Plasmodiophora	cabbage club root
Taphrina deformans	peach leaf curl
Necrotrophs	
Rhynchosporium secalis	barley leaf blotch
Helminthosporium sativum	
Septoria nodorum	
Verticillium, Fusarium	vascular wilts of many species
Facultative necrotrophs	
Lophodermella	pine needle infection
Pythium	damping off of seedlings
Armillaria mellea	honey fungus of trees and shrubs
Phytophthera infestans	potato blight

Biotrophs are generally far better adapted to their host plant, slowly draining it of nutrients but not killing it. They are able to invade healthy cells without eliciting the normal responses to infection. Most have a very narrow host range, growing in just one or a few plant species, and are unable to infect others. Many cannot grow without their living host cells and remain as spores in the soil until a suitable host is available and many will not even grow on culture media. Biotrophs usually enter plants through natural openings rather than wounds and may have highly sensitive growth responses to enable them to find their way to stomata or lenticels. Once inside, the fungus has to obtain nutrients; most have a structure called a **haustorium** as a feeding structure. The haustorium is inserted through the cell wall into a living cell, pushing between the wall and the protoplast as shown in Fig 7.6. The haustorium can alter membrane transport mechanisms to divert nutrients from the plant cell.

A necrotroph is unspecialised and may kill its host then metabolise dead tissue. A biotroph needs living cells to obtain nutrients.

Some pathogens have both biotrophic and necrotrophic features, exploiting whichever situation they find themselves in. These are described as facultative necrotrophs or biotrophs. Examples include powdery mildews and *Erysiphe graminis*.

Two diseases are described in detail below, both of which have had an important impact on people's lives.

Potato blight

Late blight of potatoes is caused by *Phytophthera infestans*, an oomycete, which is a facultative necrotroph. Potatoes at one time had no resistance to *Phytophthera* and the pathogen caused epidemics throughout Britain and Europe. Crop potatoes in warm, moist summers are particularly affected; a severe infection can cause the loss of all the leaves on a crop.

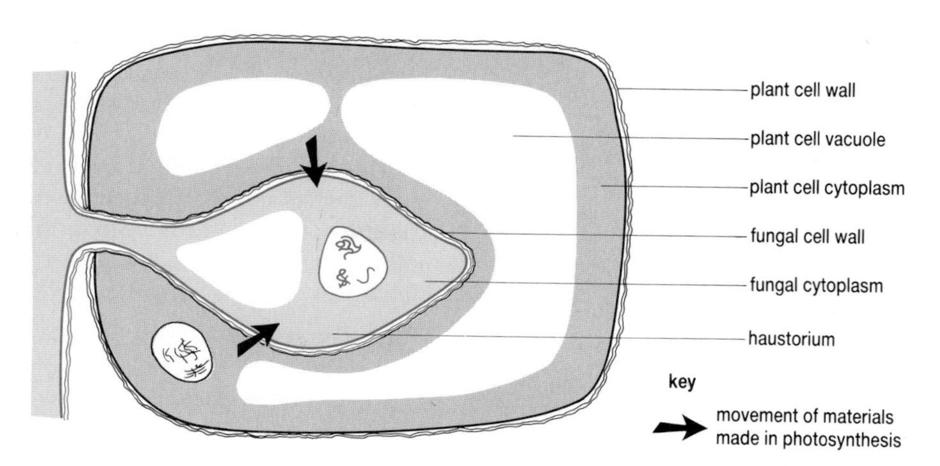

Fig 7.6 The structure of a fungal haustorium.

The disease

The pathogen is usually transmitted from infected material left over from a previous crop, or from the inadvertent planting of infected potatoes. The fungus usually infects leaves, making haustoria, but does affect tubers too. The first symptoms are a few scattered spots of infection on the leaves. The mycelium grows on the underside of the leaf and quickly affects all the above-ground parts, eventually killing them. It produces sporangia and huge numbers of asexual spores are released which are spread in the rain or by the wind to other plants in the crop. Some spores fall to the ground where they can infect tubers near the surface. Infected tubers are slightly discoloured on the outside but have a brown rot inside. Sexual spores are seldom produced in Britain as sexual reproduction requires two different mating strains, one of which is uncommon. Fig 7.7 shows the consequences of the infection. Secondary bacterial infections can cause even greater losses.

Treatment and control

The disease is much worse in wet weather than in dry. A warning about potato blight is issued in Britain by agricultural advisers when the conditions are suitable for infection, that is when the mean temperature has not dropped below 10°C and the relative humidity has stayed above 75 per cent, sometimes going over 89 per cent, for at least two consecutive days. Bordeaux mixture, which is a mix of lime, water and copper sulphate, was at one time used as a preventative spray, but modern fungicides such as the dithiocarbamates have superseded it. The pathogen is already developing resistance to some of the fungicides used. Spraying has to be repeated at regular intervals as the fungicide is washed off in the rain, and new leaves grow which are unprotected.

During the middle of this century resistant strains were bred and there are also strains which are less susceptible than others. Most early potatoes are not resistant but they are harvested before the pathogen becomes a serious pest. Good hygiene breaks the transmission chain; potatoes that were missed at harvest are removed and infected plants treated to kill the fungus before disposal. The tops of the potato plants are killed a week or two before harvesting to minimise the chances of tubers becoming infected.

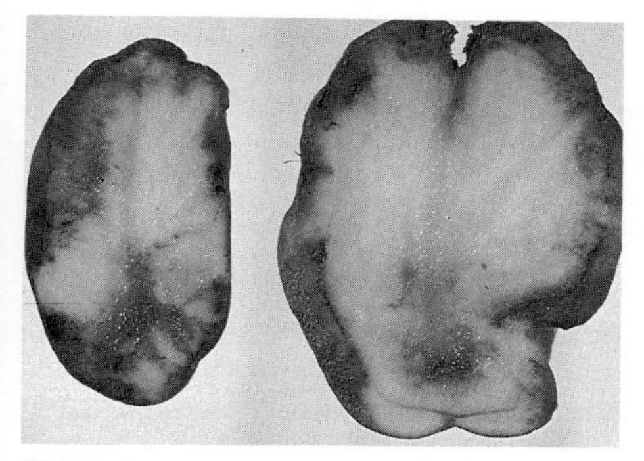

Fig 7.7 Potato tubers which are inedible because of blight infection. The outer layer appears almost normal, but there is brown rot in the middle.

Dutch elm disease

Dutch elm disease is a fungal disease of trees, particularly elm, discovered in several parts of Europe and identified in Holland. The pathogen, *Ceratocystis ulmi*, came from North America, arriving in imported timber. The timber still had its bark, hiding a species of bark beetle which carries

(a)

(b)

Fig 7.8 (a) Bark beetles leave characteristic galleries, patterning the underside of the bark of infected trees, as they tunnel through feeding on the wood.
(b) Dead elm trees were a very common sight as the infection reached its peak. The trees lost their foliage and died within one or two seasons.

the fungal spores. There was an initial outbreak in the 1930s, which eventually died down until the major epidemic in the 1970s. The second epidemic was caused by a strain more virulent than the original, which was probably imported later. It was estimated that about 11 million trees in the southern part of England had died by 1977, and in some parts of England the elm has effectively been removed from the landscape because of the disease. The fungus is necrotrophic, invading damaged areas and rapidly killing the tree. Once the tree has died the fungus continues to degrade dead tissues.

The disease

The pathogen is transmitted on the legs of a beetle, *Scolytus scolytus*, which is a sap feeder chewing into wood to get to the phloem. The beetles lay their eggs under the bark and larvae tunnel through the wood, as shown in Fig 7.8, taking fungal spores with them. The fungus invades damaged cells and grows in xylem and phloem vessels producing masses of ascospores which are carried around in the phloem. The vessels become blocked and a toxin is produced, and the plant quickly dies as a result of the accumulated damage. The spores are coated with a sticky substance that attaches them to the insects which carry the spores when they emerge as adults and fly from one tree to another.

Control and treatment

There is little that can be done to save individual trees. If the disease is recognised early enough and the trees are fairly isolated they can be injected with fungicide such as Benomyl. However, elm trees often grow close together and the bark beetles can easily fly from one to another, reinfecting treated trees. Elms do not produce much viable seed; most elms in a hedge arise by suckering from each other's roots. The trees are linked through their vascular systems and the fungus can be transmitted easily down the row. The spread, though, can be minimised by removing diseased trees promptly and controlling the beetle. Some hybrid strains of elms are more resistant to the disease and have been used to replace dead trees.

QUESTIONS

7.8 What is an appressorium?

7.9 Review the sections on biotrophs and necrotrophs and construct a table comparing the features of each.

7.10 A horticulturist grows organic fruit, that is chemicals are not used to prevent or treat infections. Give two ways that could be used to prevent fungal disease in an orchard, other than using a fungicide.

7.11 Outline a procedure you could use to investigate the effectiveness of copper sulphate mixed with ammonium carbonate solution (cheshunt compound) on the growth of a necrotrophic fungus.

7.12 Choose one fungal pathogen and construct a chart which outlines the infection cycle from a spore arriving on a plant to the next generation of spores arriving on a plant.

7.6 VIRUS DISEASES

Virus diseases are becoming more important in agriculture than fungal diseases because there are fewer control measures. New viruses are encountered as new sorts of crops are introduced, for example in the last 10 years both barley yellow mosaic virus and beet necrotic yellow vein virus have been introduced into Britain. Unlike bacteria and fungi, plant viruses are named by the plant they infect and the symptoms they produce. This can be confusing as viruses usually infect more than one host plant and

Fig 7.9 Parrot tulips owe their colour breaks to a virus infection. This variety, Estella Rynveld, starts as an almost white tulip with thin, rich red stripes on the petals, like raspberry ripple ice-cream. As the flower grows and ages the red areas increase until they become the dominant colour.

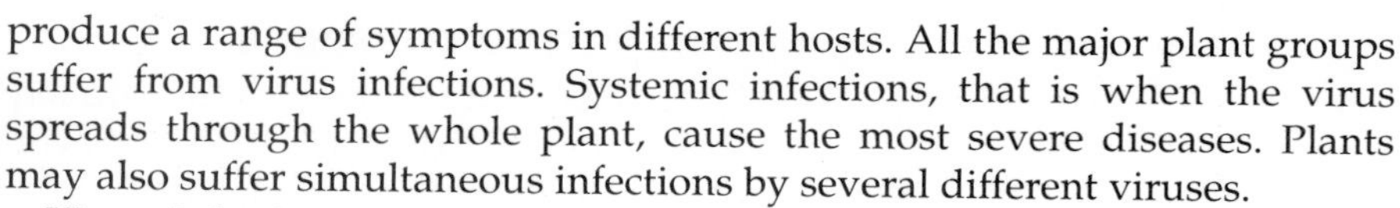

produce a range of symptoms in different hosts. All the major plant groups suffer from virus infections. Systemic infections, that is when the virus spreads through the whole plant, cause the most severe diseases. Plants may also suffer simultaneous infections by several different viruses.

Virus infections are not always easy to diagnose at first as many symptoms are like those of a mineral deficiency, or may be short-lived. An early sign is the development of paler patches by the veins of an infected leaf, later yellow spots appear. A systemic infection produces more obvious symptoms such as stunting, yellow mosaic patches, distorted leaves, pale streaks in flowers and ring marks on the leaves. Eventually a leaf may be completely yellow with the loss of chlorophyll, and cells may die leaving brown necrotic patches.

Viral activity in cells reduces normal cell metabolism and redirects the cell machinery to making more virus, which may make up a large proportion of the cell's mass. As a result the plant's growth is slowed and the yield of the crop is reduced. However, many virus infections seem to reduce the vigour of a plant rather than kill it and some have even been deliberately encouraged. A houseplant, *Abutilon*, has a 'variegated' variety which has attractive pale patches on the leaves; this is caused by a virus. Similarly the colour breaks in parrot tulips (see Fig 7.9), which have been prized for centuries, are virus-induced.

Fig 7.10 An aphid's stylets slip between cells to reach a phloem cell. In this scanning electron micrograph the stylets can be seen penetrating the epidermis of the leaf.

Vectors

Viruses cannot move independently and rely completely on being moved passively. Most are transmitted by arthropod vectors, the insects being by far the most important. There are some which are transmitted by other vectors, such as soil-living nematodes which eat plant roots or by soil fungi. Most insect transmission is by aphids, but other groups such as leaf hoppers have been found to transmit viruses too. Aphids have stylet mouthparts which penetrate the cuticle and epidermis and slip between cells into phloem vessels; Fig 7.10 shows an aphid feeding. The virus is carried into the plant with secretions such as cellulases and pectinases which help the aphid feed.

In contrast to animal viral infections, viruses are found as whole virions in plant cells, that is as nucleic acid encapsulated in a protein coat. When a vector such as an aphid feeds on the contents of a single cell it will take up whole viruses not just nucleic acid. Many viruses stick to the mouthparts and are passed on to the next plant, but others have a close relationship with the insect which remains infective for the rest of its life. The virus may even multiply in its host insect; Table 7.3 shows some of the characteristics of a close virus/vector relationship. Each virus only multiplies inside a specific host, though an aphid may be able to transmit several viruses. A few viruses are transmitted by direct contact, including tobacco mosaic, potato Y and carnation virus, which are particularly contagious.

Table 7.3 The relationship between a virus and an aphid

Non-persistant	Persistant
virus acquired in seconds	takes several hours to acquire infection
infectious immediately	infectivity latent for a few hours
infectivity lost after a few hours	infectivity for days or weeks (life)
infectivity lost on moulting	infectivity kept after moulting
virus on exterior	virus found in aphid body fluids
virus does not multiply	virus multiplies within host
not host specific	host specific

Viroid diseases

Viroids are pathogens that are like viruses but occur naturally as free RNA a few hundred bases long. Several cause plant diseases including potato spindle tuber viroid, citrus exocortis viroid and chrysanthemum stunt viroid. Plants suffering from these diseases all contain these small nucleic acid molecules that are not present in healthy plants. The molecules are unusually resistant to heat and UV light which would adversely affect ordinary RNA. Each has a region concerned with replication and a part which is concerned with the infection process, but little more is known about them as yet.

Two virus diseases are described. Each affects an important crop and is carried by an aphid vector, but control measures are different.

Potato Leaf Roll Virus

This is a particularly important disease caused by a virus transmitted by the peach-potato aphid, *Myzus persicae*. In normal potato growth the first year produces small tubers which overwinter. Farmers buy these, called seed potatoes, to grow on for a second year to produce the potato crop. It is not easy to distinguish a diseased potato at planting time without cutting it open. If infected potatoes are planted, the virus causes a great yield loss in the second year of growth.

The disease

Aphids transfer the infection from plant to plant. Infected plants may not develop symptoms for a long time, and there may be none in the first year. Eventually the leaves appear tough, the edges start to roll over onto the top part of the plant, they change colour and they lose their chlorophyll. The infection is systemic so the tubers are affected too. If infected tubers are planted or left in the soil, the disease is very noticeable the following year; in Fig 7.11 the whole plant shows leaf roll symptoms and is stunted. Carbohydrate is stored in leaves rather than translocated down to the tubers which results in fewer tubers per plant and loss of yield, upto 80 per cent in some places. The tubers are brown inside though they look normal on the outside.

Fig 7.11 Over half the leaves on this plant are showing symptoms of potato leaf roll virus infection.

Control

The disease cannot be treated so preventative measures have to be used. If aphid numbers rise they can be sprayed with insecticide or the crop can be sprayed as a preventative. Great care must be taken to remove all infected potatoes from a crop so that there are no 'strays' to act as reservoirs or overwintering quarters for the aphids. Only virus-free seed potatoes, called certified seed, should be planted. Seed potatoes are grown mainly in Scotland and Northern Ireland to reduce the risk of infection. The aphid vector overwinters in the warmer regions of southern Britain and France then migrates north during the spring and summer. As aphids dislike flying at temperatures below 20°C and in winds over 4 mph it is quite late in the year before they arrive in Scotland, when seed potatoes have already formed and are ready to be lifted. In Europe conditions are more difficult; in Holland a watch is kept on the migration and numbers of aphids, and when the numbers pass a threshold level the seed potatoes have to be dug up.

Barley yellow dwarf virus

Barley yellow dwarf virus (BYDV) is a polyhedral virus which infects over 100 species of grasses including cereal crops. A large proportion of the cereals and grasses grown in Britain are infected, giving a reduced yield. Related viruses, sugar beet yellow virus and sugar beet western yellow virus, infect sugar beet, an important crop. The virus is carried by several

Fig 7.12 An oat plant infected with barley yellow dwarf virus. Almost half of its leaves have developed a characteristic red colour instead of the normal green.

species of aphids, but the most important is *Rhopalosiphum padi*, the bird cherry aphid.

The disease

The aphid and virus have a close relationship. An aphid needs one or two days feeding on an infected plant before it acquires the virus, but it then becomes infectious for several weeks, which is effectively the rest of its life. It passes on the infection when it feeds for several hours on a healthy plant. After about three weeks the symptoms of infection appear (see Fig 7.12). Infected plants are stunted as they have fewer roots and they are a distinctive colour. Oats turn reddish purple, wheat and barley bright yellow. There may also be necrotic spots and ears of grain do not develop. Inside the plant the virus infects phloem tissue which is severely damaged.

Treatment and control

Crops can be infected as soon as the weather is warm enough to allow the aphids to move from their winter quarters, so control measures place emphasis on stopping the movement of aphids. They survive the winter as eggs on the bird cherry tree in hedges and woods, others may overwinter as adults on cereals and grasses. Aphid numbers are monitored and as soon as they start to rise they are sprayed with insecticides. Wild grasses are also infected by the virus and are reservoirs of infection. These must be weeded out of grain fields to remove the reservoir and overwintering aphids. Efforts are currently being made to breed resistance to BYDV into cereals.

QUESTIONS

7.13 Why are viruses transmitted mainly by vectors?

7.14 Outline three methods that could be used to control the spread of a virus infection in a potato crop.

7.15 Very briefly explain why virus diseases make leaves lose their green colour.

7.7 BACTERIAL DISEASES

There are some very damaging bacterial diseases, but on the whole bacteria are less important than fungi and viruses. The first to be given much attention was fire blight, an American disease of pears which was investigated last century. The soft rots caused by *Erwinia* species are important from the consumers' point of view as they are pests of stored potatoes and onions.

Bacteria enter the plant through stomata and lenticels, or through wounds. Young dividing cells with a high nutrient content are the most likely to be infected but other tissues may also be damaged. Once inside they secrete enzymes such as pectinases which degrade the tissues, which they can then colonise. Necrotic spots on leaves and stems grow outwards as the bacteria multiply. Some grow in the phloem and xylem and are spread to other parts of the plant in the sap streams. Bacterial capsules and secretions may reduce transport through the affected tissues causing wilting. Other bacteria, particularly *Agrobacterium tumifaciens*, stimulate unusual cell proliferation resulting in galls and tumours.

The two examples discussed below are not great problems in Britain, but one, fire blight, was the first disease (animal or plant) proved to be caused by a bacterium, and it is still a major problem to fruit growers in many parts of the world. The other, crown gall, is interesting because its causative bacterium, *Agrobacterium tumifaciens*, is important in genetic engineering for carrying genes into plant cells.

Fig 7.13 Shoots which are infected with fire blight wilt, then the leaves start to look scorched and the shoot dies. One shoot on this tree is affected, but the surrounding branches will develop symptons before long.

Table 7.4 Some bacterial pathogens

Pathogen	Host	Disease
Agrobacterium tumifaciens	fruit trees	stem galls, loss of yield
Pseudomonas tabaci	tobacco	wild fire, wet rot in leaves or necrotic spots
Xanthomonas oryzae	rice	leaf blight
Erwinia chrysanthemi	chrysanthemums	vascular wilt

Fire blight

Fire blight is caused by *Erwinia amylovora* and infects a range of shrubs as well as apples and pears. It was first isolated in North America but has spread to Europe and the East. It appeared in Britain in the 1950s, becoming a notifiable disease. Despite strict control measures it is widespread, but cool British springs keep infection at a fairly low level. Late flowering species such as Laxton's Superb suffer particularly and may not be grown in some areas. Infected leaves stay on the tree making it look as though it has been scorched in a fire (see Fig 7.13).

The disease

Bacteria enter through stomata, the receptacle, wounds, and parts with a thin cuticle such as nectaries. They secrete a toxin which brings about rapid wilting and necrosis of the tissues. The flowers and shoots wilt, turn brown and die rapidly. Bacteria spread down the shoot and into the branch, then into other branches, and the tree eventually dies. Bacteria ooze out of lesions, or cankers, on the bark, particularly in damp weather. The bacteria are spread to other trees by rainwater splashes and visiting insects. The cankers become inactive in the winter but produce lots of bacteria in the spring as the temperature goes up.

Treatment and control

There is no treatment except to destroy diseased trees.

Fig 7.14 Crown galls on beets result in some odd shapes.

Crown gall

Crown galls are caused by *Agrobacterium tumifaciens*, which is a soil inhabitant. It can infect over 1000 species of plants, but does not usually kill them. It causes serious crop losses in fruit trees and beet, and kills hops.

The disease

The bacteria enter through wounds, then stimulate host cells to multiply rapidly and make large lumps called galls. Different species of plant produce different shapes of galls; Fig 7.14 shows a crown gall. The bacterium does this with a plasmid, called the T_i plasmid, which becomes integrated into the host cell chromosome. The plasmid carries genes that cause the host cell to divide and to make amino acids called opines, not normally made by plant cells. The bacterium uses the opines in its growth.

Treatment and control

There is no treatment, though some antibiotics kill the bacteria, so control rests in using clean stock and the careful disposal of infected material. Any wounds, such as grafting and pruning wounds on fruit trees, should be protected against infections. Crop rotation helps to limit the spread of the disease as it removes susceptible plants.

QUESTIONS

7.16 Review section 7.2 and explain why bacteria are not very serious plant pathogens.

7.17 Which are the two main entry points for bacterial infections?

7.8 CONTROL OF DISEASE BY BETTER PRACTICE

The best way of controlling plant diseases in agriculture is by the management of pathogens. No one method is entirely relied upon, but a combination of different approaches is used to minimise infections and to eradicate sources of infection.

Hygiene

Improved hygiene is the only realistic method of control of most pathogens. Though there are cheap effective fungicides there are no cheap anti-viral chemicals and disease control has to centre on prevention. All sources of infection must be removed, and any material from an infected crop must be disposed of, usually by incineration. Any wild host plants must be controlled so that they do not act as reservoirs of infection. Weed hosts are eliminated by the traditional weeding or by the use of selective herbicides; these are chemicals which will kill the offending weed but not the crop plant. Any stray plants of a previous crop which escaped harvest can be removed in the same way.

Crop rotation, which involves growing crops on a different piece of land each year over a cycle of years, deprives soil-borne pathogens of 'victims'. Spores and other resistant structures may not survive until the rotation brings vulnerable plants to that area again. There are some diseases, for example heart rotting fungi, that spread from tree stumps to nearby trees. These can be discouraged by a form of biological control. Foresters can introduce a harmless fungal species which only degrades dead wood into a tree stump or a logging wound. This acts as a commensal and prevents the more harmful species from establishing itself. This is one of the few biological control methods available at the moment.

Good hygiene requires the removal of all infected wild and crop plant material before a crop is sown.

Soil sterilisation is impractical on a large scale, but it can be done on a small scale. Sterile soil enables seedlings to get off to a good start; once the plantlets are well-grown they can be transplanted. When sources of infection have been eliminated, a fresh start can be made with healthy stock.

Virus-free stock

If a crop is to be kept clear of virus infection the starting material must be virus-free or treated to eliminate virus infections. Few viruses are transmitted through seeds or pollen so seeds can be used to ensure virus-free plants. Many plants carry infections, and for many years it was almost impossible to obtain plants of some species without a virus infection. If a virus-infected plant is used as a source of cuttings in a nursery then all the progeny plants will carry virus too, though if the virus does not infect all the parts of a plant there may be some which are infection-free. **Grafting**, which is propagation by taking sections of a plant and transplanting onto another plant, will also transmit systemic viruses.

A treatment was developed in the 1930s which could eliminate some viruses from infected stock. Plants were held at high temperatures for a period of time. This was minutes for some viruses, weeks for others. For instance, potato tubers infected with potato leaf roll virus were held at 37.5°C for 25 days. This is still the only way of curing most virus infections but it has fallen into disuse as there are now better methods of obtaining virus-free stock. Virus-free plants can now be produced using the micro-propagation techniques described in Chapter 5, and these are now routine for some important crops. Many virus infections do not reach the growing points where new cells are produced, the **meristems**, or if virus reaches them it is at low levels and individual cells may be disease-free. The meristems can be heat-treated if necessary to remove any viruses. These are then used as propagating material and callus cells can be used to produce large numbers of plantlets which are free of virus. Samples of the virus-free tissue can also be taken and kept at low temperatures for use at a later date.

Fig 7.15 The windows and ventilators of this greenhouse are fitted with mesh screens. These bar the passage of aphids and enable the grower to produce plants free from viruses carried by aphids.

Growing different varieties

Monoculture encourages the development of pathogens which can overcome resistance mechanisms. The more a particular cultivar of a crop is grown, the more likely resistance will be lost. It is better to have blocks of different cultivars grown in the same area which limits pathogen spread and eases selective pressures. Similarly it is better to grow different varieties in the autumn and spring as this again helps control the pathogen.

Improving induced resistance

Manipulating the ability of a plant to make a phytoalexin is one possible way of protecting a plant from infection, though there are few examples. A systemic, non-fungicidal chemical known as DCCP is used to protect rice from blast fungus leaf infection. Plants treated with the compound produce two phytoalexins in much greater quantities than untreated plants. The plants then respond to infection with the blast fungus by hypersensitivity and inhibition of fungal growth. Another possible approach is to develop synthetic analogues of phytoalexins to use as treatments for infections. Iron also seems to boost resistance to plant pathogens but the mechanism is not fully understood. To date, there is very little that parallels the stimulation of the immune system in animals.

Vector control

The most important disease-carrying vector in Britain and much of Europe is the peach-potato aphid which transfers many viruses. The aphid is around all year, but is restricted to mild areas in winter. It breeds prolifically during the spring and early summer, expanding northwards. As most vectors are insects, insecticides play a large role in control but there are drawbacks to these as a sole means of control. Many aphids and other vectors become resistant and other, valuable, insects may not, so there is an inevitable toll on other inhabitants of ecosystems, and insecticide levels may accumulate higher in the food chains. Some insecticides are more selective in their action than others, but even so spraying has to be done at times of the day when insect pollinators are not active. In glasshouses other techniques can be used. The simplest is shown in Fig 7.15, in which mesh which is too small for aphids to get through is suspended over all doors and ventilators.

Winter is a difficult time for insects; the cold and lack of vegetation kills large numbers. Aphids use particular plants as winter refuges, for example *Myzus persicae* overwinters on a range of plants including the Prunus family – these are the cherries and their relatives. Particular shrubs and other vegetation harbouring species of pests may have to be removed from crop areas.

Insect vectors can be controlled by insecticides, removal of refuges and by avoiding times when insects are prolific.

In the past a piece of ground would be used for one crop a year, planted in spring and harvested in late summer, but now there are crops which can be planted at other times of the year with staggered harvest times. As a result there are aphid food plants around in late autumn and early spring when there didn't use to be. Aphids live on these crops, which are sometimes described as 'green bridges', and infect spring crops. Though one farmer can coordinate activities to avoid green bridges, several farmers may need to work together to prevent overlaps between different farms.

Biological and chemical control measures

A recent innovation is to use biological control of the vector, but there are few examples at the moment. Cultures of bacteria which parasitise insects can be released into a glasshouse to control the numbers. *Bacillus thuringensis* has been bred on a large scale specifically for this purpose, and is used on a commercial scale. The bacterium produces a toxic protein which kills caterpillars of some moths and butterflies, but unfortunately these are not transmitters of disease though they do cause damage.

Other ideas are being investigated but will not be available for many years. One is to engineer genes into plants that enable the plant cells to make insecticidal proteins; another is to engineer genes for enzymes such as chitinase into soil bacteria which would affect opportunist, soil-borne fungal pathogens. Generally little research has been done on the biological control of fungi as there are so many cheap fungicides available.

Chemical dusts, smokes and sprays

The only effective chemical controls are fungicides, apart from sterilising solutions to treat soil and some disinfectants which can be used on individual plants. Soil pathogens can be killed by using a number of chemicals, such as formalin, on the soil. Unfortunately this is very expensive and most of the chemicals are are not specific, and so it cannot be done on a large scale. It may, in any case, be counterproductive as it also kills the non-pathogenic organisms such as nitrogen fixers and decay organisms responsible for soil structure. Fungicides are useful to treat soil which is to be used to germinate plants for later planting out, particularly to kill damping-off fungi such as *Pythium*.

Fungicides do not have to be applied until there are enough of the pathogen present and conditions are suitable for the infection to get started. Though fungi produce large numbers of spores many are inactivated, for example enough ultraviolet light or dessication will kill them. Viable spores only germinate when conditions are suitable for growth, that is when it is warm and moist, and this is when fungicides are most effective. Weather conditions are monitored and warnings given when there is a danger of fungal infection. If there is an isolated outbreak of disease affecting only a small portion of a crop it may not be cost-effective to spray against the infection. A scale of disease incidence can be used to decide when it is most economical to spray a crop.

Multi-site fungicides affect a range of metabolic processes; single-site fungicides affect only one or two.

Fungicides may affect one or two metabolic activities in a pathogen, called **single-site fungicides**; others affect a wide range of activities and are called **multi-site fungicides**. The multi-site compounds often contain copper which upsets energy production. Most systemic fungicides are single-site, for example Benomyl, and some of the organo-sulphur compounds affect DNA synthesis. These compounds enter cells and are altered to make substances that inhibit mitosis in the fungus. Others may affect protein synthesis or synthesis of components of the cell membrane or wall. It is relatively easy for a pathogen to evolve resistance to single-site fungicides but much harder to become resistant to multi-site ones.

Fungicides can be used as dusts for seed and root dressings or for whole plants. A problem with dusts is that they drift with air currents, particularly if applied over a large area, and may contaminate other crops. Many fungicides are formulated as wettable powders that will make a solution or suspension which can be sprayed onto plants. As plants have water-repellent cuticles, sprays may have another chemical added to help the fungicide stick to the plant. They can also be made into granules or smokes for dealing with special areas, for example a smoke could be used to fumigate a greenhouse. New fungicides are being sought which can be sprayed onto the top of a plant and will then be translocated down into the roots to deal with soil-borne pathogens.

If fungicides are likely to be used several times in a season, it is better to use two together to minimise the development of resistance in the pathogen. The fungicides should be multi-site rather than the same single-sites all season. All fungicides should be used with care and only when they are likely to have the best effect. Some fungicides cannot be applied to crop plants close to the time of harvest as they leave residues on the crop. Many fungicides need extra care (see Fig 7.16), because they are hazardous to the people working with them; there are rigorous rules for the safe handling and storage of pesticides of all types.

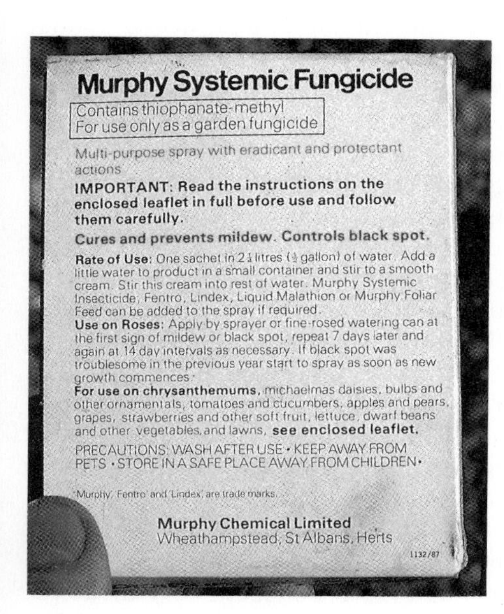

Fig 7.16 Fungicides and other control agents are packaged with clear instructions for safe use. People may need special training and equipment before they carry out certain large-scale agriculture treatments.

7.9 BREEDING RESISTANT PLANTS

A huge amount of effort is devoted to breeding strains of plants that are resistant to disease. Many modern strains of crops lack the range of resistance mechanisms to pathogens that their antecedents had, though they may have features which are better in other ways. The plants in Fig 7.17 show how different a wild plant and the 'domesticated version' may be. The process of breeding has meant that resistance as well as other characteristics may have been lost in the genetic reshuffle that occurs whenever pollen and ovules are formed.

(a)

(b)

Fig 7.17 Selective breeding for generations has resulted in modern fruits and vegetables which bear little resemblance to their wild ancestors: **(a)** shows a modern pineapple fruit and **(b)** the wild pineapple.

Resistance to pathogens is usually carried genetically. A particular feature such as a thick cuticle is due to the possession of one or a few genes which cause the plant to make a thick wax layer; other features, such as the ability to make phytoalexins, are also genetic. A wild population of plants is really a mixture of genetically different individuals, and particular plants will have different combinations of resistance genes.

The presence of a pathogen in a crop or wild plant population results in selection for plants which have resistance features; those which do not have these features are infected and struggle to survive or often die. The populations of pathogens are also genetically different individuals. Each carries a different combination of genes, some of which will help it invade its food source. If an individual pathogen attempts to invade a plant it will succeed if the plant does not have the appropriate resistance genes. When the plant does have the right resistance genes the pathogen cannot infect and will die. A particular plant may be resistant to one strain of pathogen but susceptible to another strain with different invasive genes.

Plants and pathogens develop resistance genes and infection genes together. The ability of a pathogen to cause an infection depends on its invasive genes and the match with the host's resistance genes.

When a resistant cultivar of a crop is first introduced, the pathogen cannot infect it. In due course, however, the pathogen will evolve a mechanism which enables it to overcome the resistance. For example, if resistance is due to an inhibitory chemical the plant makes in its cuticle, any pathogen which can make a degrading enzyme, can detoxify it or has an alternative pathway not affected by the toxic chemical will be able to grow in the plant. In this way, plants and pathogens evolve defensive and invasive mechanisms in response to each other. As new resistance genes arise the pathogens evolve new virulence genes. Often resistance to disease may not be a straightforward block of the progress of an infection; it may be due to features which allow the plant to tolerate or survive the infection better.

The breeders' targets

Plant breeders have definite targets in their search for resistant strains. An ideal strain will have more than one resistance gene which would make it more difficult for a pathogen to overcome the resistance. Problems arise when strains have got some of the desired resistance characters but not all of them, since the pressure to put them on the market early may be very great. For example, tomatoes have had three resistance factors identified, which confer resistance to TMV. These genes are being engineered into a strain to produce a very resistant tomato, however tomatoes with just one of these have been grown and pathogen resistance has already been reported.

Knowing about the infection process gives us clues as to what to do to obtain better strains of plants. For example, bean rust on haricot bean (*Phaseolus vulgaris*) is caused by a fungus whose spores germinate on the leaf surface. A hypha grows directly to a stomata where an infection peg develops and enters the plant. It has been discovered that the key to the infection process is the shape and size of the microscopic ridges on the surface of the leaf. The hyphae form infection pegs when they have grown over a ridge as high as a stomatal guard cell. If plants could be bred with smaller ridges the infection pegs would not develop and the plant would be resistant to the fungus.

Ideally, more than one resistant strain should be bred so that farmers in an area can grow different versions and minimise the evolution of pathogen resistance. However, all farmers would want the strains which gave the best possible yield, so all the strains would have to be bred for the same growing characteristics so that no-one was disadvantaged by growing one strain rather than another.

Breeding techniques

Many techniques are used to try to breed resistance into crop plants. Traditionally, strains are crossed using pollen transfer from one plant with the desired disease resistance to another with good growing characteristics, and collecting the seed produced. This is grown and the performance of the individual plants assessed. Good possibilities are bred on or crossed again until, several years and hundreds of thousands of plants later, a strain is produced which has the desired characteristics. Unfortunately desirable characteristics are easily lost during meiosis, so many seasons of crop improvement work can produce thousands of plants which are no better, or even worse, than the originals.

Micropropagation offers a short cut to crop improvement. The cells in a protoplast culture are each capable of regenerating a plant, however the plants produced vary in their characteristics though they all carry the same genes. Regeneration is quick compared to sowing seeds and waiting for the next generation, and lots of different plants are produced. These can be checked and selected or discarded rapidly. Once a desirable strain has been produced and tested it can be propagated very quickly, shortening considerably the time taken from trials in the research institute to the improved strains being available to the farmer.

Pollen culture

Pollen cultures are used to produce pure-breeding or homozygous plants whose characteristics are easily determined. This can take years using conventional techniques but pollen culture reduces the time to just a couple of years. If the plants carry resistance genes they can be a source of genes to be bred into crop varieties. Inside a pollen grain male gametic nuclei are produced which would, in due course, be used to fertilise an ovule. The nuclei are haploid, that is they have only one set of chromosomes carrying one set of genes. The pollen grains can be cultured on tissue culture medium with regulators to stimulate root, then shoot, formation. The plantlets are haploid and can be treated with colchicine to bring about chromosome doubling. This results in flowers which can be self-pollinated to produce plants which are diploid and homozygous for each character. Some pollen grains undergo spontaneous doubling producing diploid plants directly.

The techniques of genetic engineering are also being applied to crop resistance. It is already possible to incorporate genes into plant cells using a plasmid carried by *Agrobacterium tumifaciens*. So far genes have been transferred into broad-leaved plants, but there has been less success with the cereals and grasses. This may offer a quick way of getting desirable genes into a crop plant.

QUESTIONS

7.18 Potato leaf roll virus was a problem in the USA, but was much less of one in India until seed potatoes were routinely stored in cold storage instead of at normal temperatures. Read the section on virus-free stock and suggest why this could be.

7.19 Explain what the terms 'boom and bust' and 'green bridge' mean in plant disease.

7.20 A farmer discovered that 40 per cent of his potato crop had a virus infection which made the inside of the tubers brown and inedible. As he could not sell his potatoes he ploughed them into the ground in disgust. He thought that he could at least use them as organic material in the soil to fertilise it. He replanted with new seed potatoes the following year which he bought as certified virus-free stock. To his horror he found that when he lifted his second year's crop the potatoes were infected again. Explain three ways in which his virus-free potatoes could have been infected.

7.21 You have been provided with a diseased plant and several healthy ones of the same variety. You also have a pure culture of a fungus which you suspect is causing the disease. What investigations could you carry out to show whether or not the fungus you have is the cause of the plant disease?

7.22 Give three ways in which a farmer could ensure that his crop of glasshouse tomatoes remained free of TMV.

SUMMARY

Simple chemical treatments of lime and sulphur have been used to treat plant disease for centuries. The main plant pathogens are fungi and viruses, and these cause a wide range of infections spreading from continent to continent and reducing the yield of major crops. Plant pathogens are easily spread to non-endemic regions where the native plant populations have no resistance. Agricultural practices can encourage the spread of disease and encourage the development of even more invasive species.

Pathogens damage plant cells, reduce their photosynthetic capacity and drain nutrient reserves. Fungal parasites, which may be unspecialised necrotrophs or specialised biotrophs, can be controlled using fungicides. Viruses are usually controlled by hygiene and vector control. Resistant strains of plants are now being bred to combat infection by pathogens.

Chapter 8

HUMAN DISEASE

LEARNING OBJECTIVES

At the end of this chapter you should be able to:

1. appreciate how the incidence of disease can be reduced;
2. understand the causes of a range of diseases;
3. give detailed accounts of selected diseases and their control.

8.1 INTRODUCTION

In this chapter the focus is on disease in one particular mammal, ourselves, but our diseases are not very different from those suffered by other mammals, though the severity of the diseases varies. We are afflicted by pathogens from all microbial groups, particularly bacteria and viruses. Protozoa cause few diseases and serious fungal diseases are rare. Each kind of microorganism is considered separately, with a brief survey of the problems they cause, and a few serious diseases are covered in more detail. One group of pathogens is not included. These are the parasitic worms and flukes, which though they cause major diseases are beyond the scope of this book. Before working on this chapter check your syllabus to see which diseases you need to study in depth.

8.2 COMMUNITY HEALTH

Environmental management and vaccination reduce the incidence of infectious diseases.

In Britain the incidence of infectious disease, that is the number of people catching a disease, has dropped substantially this century, mainly through careful management of our environment. People's resistance to infection has been raised by general good health, by being better fed and by having adequate housing. People can be protected from specific diseases by vaccinations. If someone does contract an infection there are therapeutic treatments with drugs and antibiotics which are available for most diseases. These have completely changed the mortality statistics so that infections which were once major killers such as diphtheria are now rare in countries where medical help is readily available. Fig 8.1 shows the pattern of mortality from diphtheria during the first part of this century before antibiotics were discovered.

Though our bodies are protected by natural barriers and immune system activity, we can take many other measures to reduce the chances of us, or our livestock, contracting infections. Community hygiene can be practised, for example ensuring a safe water supply and satisfactory waste disposal. Trained staff working in primary health care in clinics and visiting homes can encourage people to adopt hygienic practices which reduce disease. Families with young children may be supported financially so that the children can be provided with vitamins and minerals needed for healthy growth and disease resistance. A programme of vaccinations can be organised to protect vulnerable individuals. Those who do succumb to infections are cared for and treated, and primary health care clinics are backed up by hospitals which undertake specialist treatments. Careful monitoring of disease patterns and isolating sources of infection help prevent the spread of infectious diseases.

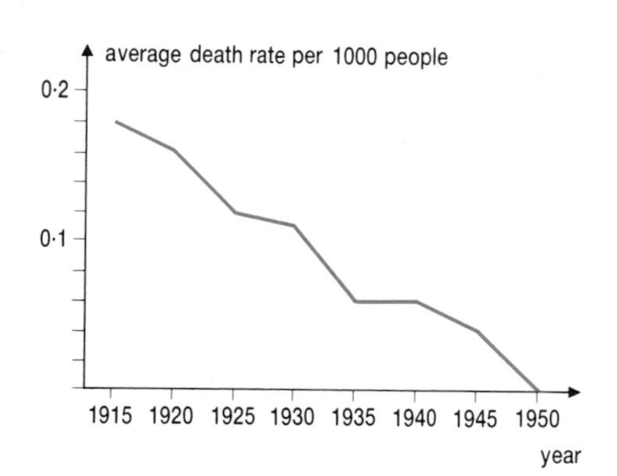

Note: the death rates are a five-yearly average.
From: Report of Medical Officer for Birmingham, 1952.

Fig 8.1 Death rate from diphtheria in Birmingham 1914-52

A community has a network of support services which its members can draw on and which benefit the whole community; a summary of the support services in Britain is given in Fig 8.2. The development of simple diagnostic kits and cheaper generic medicines, rather than expensive formulations with a trademark, mean that health care can be taken to the places that need it most – the remote or developing areas.

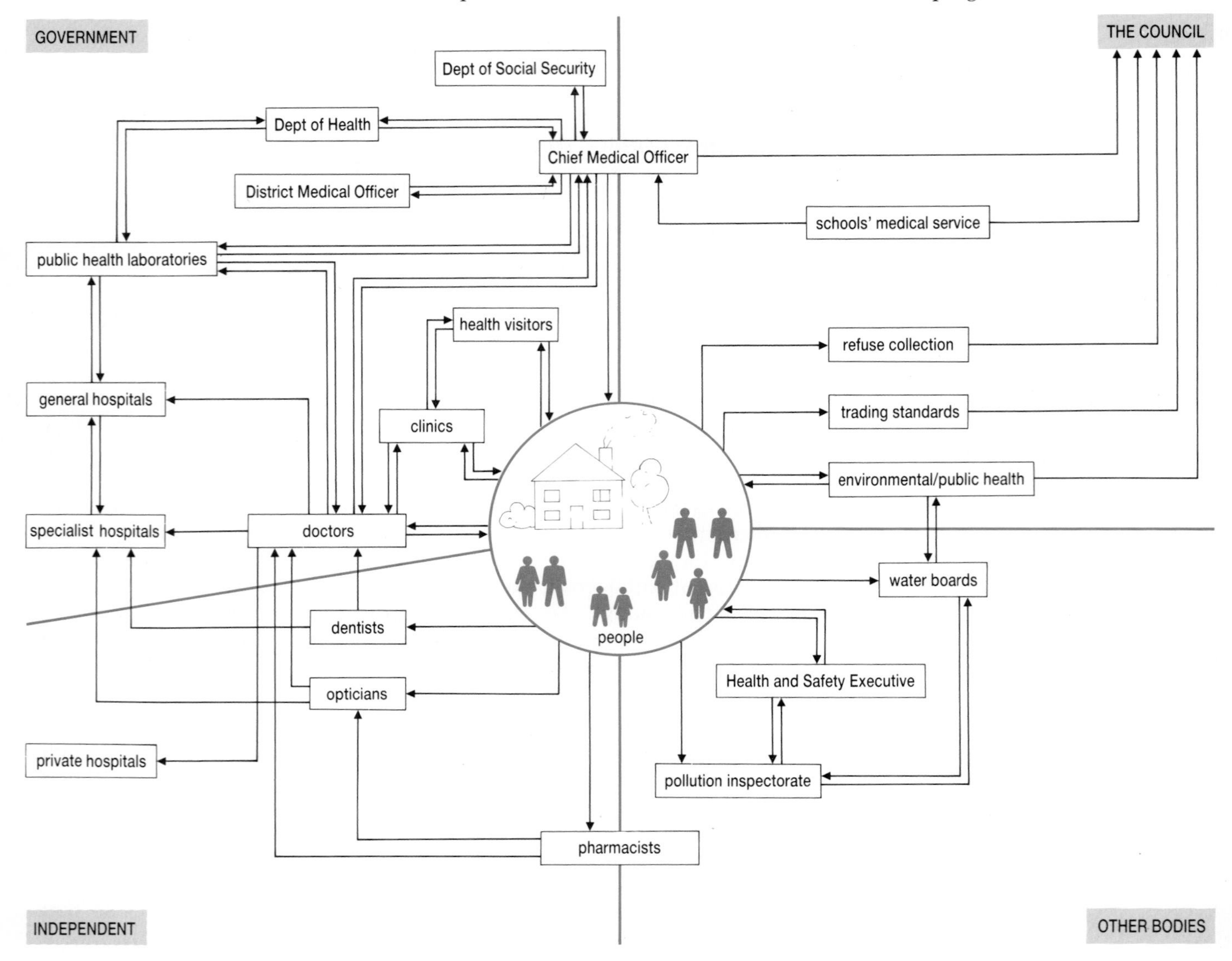

Fig 8.2 The health web in Britain.

8.3 SOURCES OF INFECTION

Pathogens come from infected individuals or from other reservoirs of infection. There are three main reservoirs of human infection which have to be dealt with in any effective health policy:

1. Individuals who have had an infection but who are not completely clear of the organisms and become **symptomless carriers**. They excrete the pathogen but do not show any signs of infection. These have to be identified and treated to clear the infection if possible.

2. Some animals harbour pathogens; a disease caused by a pathogen which infects both humans and animals is called a **zoonosis**. We share many diseases with other mammals, for example bovine TB and sleeping sickness. The pathogen has to be dealt with in both animals and humans for effective disease control. This is particularly difficult if

wild, as opposed to domesticated, animals are reservoirs. Some important zoonoses are listed in Table 8.1.

3. Many pathogens are found in **soil and water**, and infections can be contracted by contact with these. Water-borne pathogens are the easier to control with water purification; the control of soil-borne organisms is much more difficult.

Table 8.1 Animals which act as reservoirs of infection.

Disease	Organism	Reservoir
brucellosis	*Brucella abortus*	cows
plague	*Yersinia pestis*	rats and other rodents
rabies	Rhabdovirus	dogs and related mammals
sleeping sickness	*Trypanosomas brucei*	bushbuck and other large antelope
toxoplasmosis	*Toxoplasma gondii*	dogs
yellow fever	Togavirus	monkeys

THE ERADICATION OF SMALLPOX

There have been some impressive successes with the eradication of diseases from particular areas, and one spectacular example of the elimination of a disease completely. Fig 8.3 shows Ali Maow Maalin from Somalia. A remarkable person, he was the world's last case of smallpox. He developed the disease in October 1977 but recovered and is now a primary health care worker trying to reduce the incidence of other diseases such as measles which killed his younger sister. Smallpox was common in the Old World for hundreds of years; many thousands died and others were permanently scarred. The virus causes an extensive skin infection which can even spread to the lining of the gut. It is extremely infectious and killed millions of Incas and native Americans when it was taken to the New World with early settlers. The disease needs intensive care and patients have to be isolated when nursed to prevent the disease from spreading.

Smallpox was controlled by a vaccine which was very effective at protecting large populations so the incidence declined as vaccination became more common. The virus only infects people, so there were few reservoirs of infection. A decision was taken by the World Health Organisation about 20 years ago to focus efforts on a 10 year campaign to eradicate the disease by vaccination, and WHO began tracking down infected people. At that time they estimated that there were about 10-15 million cases of smallpox occurring each year. Health workers worldwide visited communities, no matter how remote, and administered vaccine. Any people found to be suffering were treated and the spread of infection was blocked. The last few cases were in Africa, in 1977, where just over 3000 people were diagnosed as having smallpox, five in Kenya and the rest in Somalia. Despite a combination of heavy rain and civil war, health workers were able to track down the last few cases. It had taken 10 years, 9 months and 26 days and about $200 million, which was far less than the amount spent annually on controlling and coping with the disease. Every case of chickenpox or other skin rash was examined to ensure that there were no other smallpox sufferers with a minor infection. The eradication was certified by the World Health Authority in May 1980 and the practice of routine vaccination of the disease was stopped to reduce the amount of virus in laboratories; only military personnel in a few countries are still vaccinated.

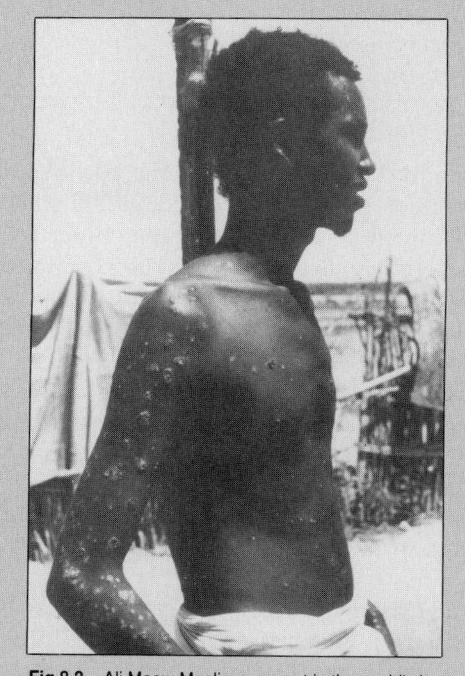

Fig 8.3 Ali Maow Maalin came out in the world's last smallpox spots on 26th October 1977.

QUESTIONS

8.1 What are meant by the terms 'zoonosis' and 'carrier'?

8.2 List three measures which can be taken by a community to ensure the health of its members.

8.3 Examine Fig 8.1. Why are the death rates quoted as per 1000? Comment on the factors which may have lead to the decline of diphtheria.

8.4 BACTERIAL DISEASES

Bacteria cause a large number of diseases which used to be almost impossible to treat. Now most are curable, given the appropriate medical resources and access to trained staff in the early stages of infection. Mycoplasmas, a group of bacteria which do not have cell walls, cause troublesome infections which are hard to clear up and are implicated in a number of diseases whose causative agent is unknown. Most groups of bacteria have pathogenic members which colonise and exploit every part of the human anatomy. Table 8.2 summarises the bacterial pathogens which you have met in the previous chapters.

Table 8.2 Summary of bacterial diseases mentioned in the book and their causative organisms

Disease	Organism
diphtheria	*Corynebacterium diphtheriae*
TB	*Mycobacterium tuberculosis*
leprosy	*Mycobacterium leprae*
anthrax	*Bacillus anthracis*
tetanus	*Clostridium tetani*
whooping cough	*Bordetella pertussis*
brucellosis	*Brucella* spp
bubonic plague	*Yersinia pestis*
cholera	*Vibrio cholerae*
syphilis	*Treponema pallidum*
Rocky Mountain spotted fever	*Rickettsia rickettsii*

In a developed society we are more likely to encounter some pathogens than others, not necessarily the ones which cause concern on a global scale. In developing areas even quite trivial infections can be life-threatening. The most likely pathogens to be encountered in Britain are streptococci and staphylococci, which are opportunist infections, and the enteric bacteria. Most of these are common commensals and, with the exception of some enteric bacteria, are not usually controlled by vaccination.

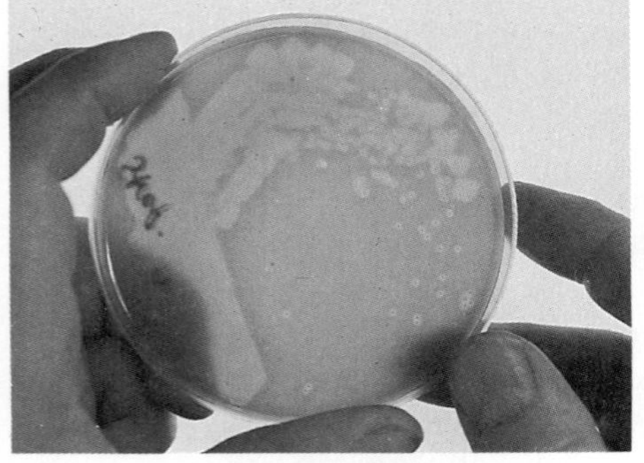

Fig 8.4 Haemolytic streptococci growing on agar containing blood have broken down the red pigment, haemoglobin, around the colonies.

Some **streptococci** are particularly unpleasant, especially *Streptococcus pyogenes,* which has a toxin affecting blood cells and causes pus-forming infections. Many people encounter it as 'strep throat', a form of sore throat: Fig 8.4 shows a culture of these bacteria; the clear zone around the colonies is due to haemolytic activity. Other streptococci cause bacterial pneumonia, meningitis, scarlet fever and impetigo, a skin infection, but these can be treated with penicillins.

Staphylococcal infections also afflict many people. The main culprit is *Staphylococcus pyogenes aureus* a commensal. This causes pus-forming infections too, such as boils and abscesses, and it also causes impetigo. Its toxin also causes food poisoning if it passes from a septic infection on someone's hand into food.

Diarrhoeal diseases

Everyone has had the misfortune to suffer an infection causing diarrhoea, whether from eating a take-away meal or 'holiday tummy'. For healthy people these infections may mean an uncomfortable and embarrassing day or so and an unexpected inch off the waistline; for young children or malnourished people diarrhoea is life-threatening.

The infection known generally as food poisoning can be caused by a number of bacterial species in the genus *Salmonella* as well as *Escherichia coli* and other bacteria. There are also more serious infections affecting the gastrointestinal tract such as cholera, dysentery and typhoid, all of which are serious epidemic diseases. Acute diarrhoeal diseases are the main cause of death in young children worldwide, at about five million deaths each year.

Salmonella food poisoning

The organism

Food poisoning is caused mainly by *Salmonella* bacteria, though *Shigella* and *Campylobacter* species also account for a number of cases. The most common culprit is *Salmonella typhimurium*, but others such as *S. enteriditis* and *S. virchow* are often recorded. Together they account for half the food poisoning cases reported in Britain. They are flagellate, rod-shaped bacteria, shown in Fig 8.5(a), which are found in the gut flora of many farm animals, especially those reared intensively with highly concentrated feeds. Most chickens in deep litter houses have salmonella in their gut. An increasing number of salmonella are finding their way into unpasteurised milk and eggs, and hence into foods made with raw eggs.

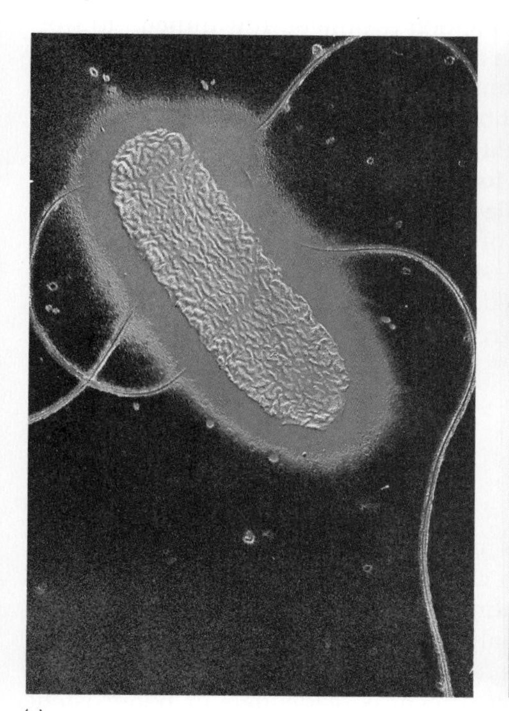

(a)

(b)

Fig 8.5 **(a)** A *Salmonella enteriditis* bacterium, with flagella. This species is causing an increasing number of cases of food poisoning.
(b) Chickens are reared intensively in many different systems. In a deep litter house the birds are free to move, to reach their food in hanging tube feeders, but droppings are cleared out very infrequently.

The disease

The bacteria are not very virulent so about 10 million live bacteria have to be eaten before an infection starts. In farm animals the bacteria are confined to the gut; meat and poultry become contaminated after slaughter. As meat is chilled or frozen straight after slaughter there is little bacterial growth until defrosting. Bacteria can pass from defrosting meat in a fridge or a display to other foods, particularly if there is a film of moisture to swim in. People handling chicken, for example, can transmit the organisms to other items on their hands or on utensils. The most common source is meat which has not been properly cooked and left in warm surroundings or food which has come into contact with bacteria.

Symptoms begin within a few hours though it may take a couple of days. The patient has pains in the abdomen, feels feverish, suffers diarrhoea and may be sick. This can go on for several days, resulting in an enormous loss of fluid and electrolytes from the body. In malnourished children, elderly people and people with other infections as well the fluid loss can be fatal. The bacteria colonise the small intestine and enter the cells lining the gut. They multiply within these cells, killing them and provoking an inflammatory response. The damage to the microvilli on the intestinal mucosa leads to absorption problems, and possibly osmotic problems and problems with digestive enzyme production.

Treatment and control

Antibiotics are not very effective against these diseases as it is difficult to get enough into the cells where the bacteria are causing damage. The main treatment is care of the patient, supplemented with fluid replacement in severe cases, that is glucose solution containing electrolytes administered as a drip into a vein. Most patients recover within a week.

The disease is controlled by improving the standards of hygiene at farms and in all stages of food handling and preparation. In some countries bacteria in carcasses are killed by radiation treatment. It is difficult to stop chickens from acquiring the bacteria as they are found in faeces in the deep litter in chicken houses and in contaminated feed. Some chickens develop a carrier state and can infect others. It has been suggested that a good flora of commensals in a chick's gut might reduce the ability of salmonella to colonise, so efforts are being made to find a way to get 'good' commensals into the chick's gut before salmonella species can get in.

Food handlers have to be aware of the hazard posed by uncooked meat, however if foods are heated to at least 56°C the bacteria will be killed. Food should always be adequately cooked and extra care should be taken when cooking poultry which has been frozen.

Typhoid

The organism

Typhoid is a much more serious disease caused by another flagellate salmonella, *Salmonella typhi*, which can only infect humans. It is excreted via the gut and is transmitted by eating food or drinking water carrying the pathogen. It is a virulent species; only a few organisms are needed to start an infection. The organism is endemic in many parts of the world, particularly in areas where water treatment is inadequate or non-existent and drinking water is polluted with raw sewage.

The disease

The bacteria survive the stomach and multiply in the small intestine. They make their way through the cells lining the intestine into lymphoid tissue such as lymph nodes and the spleen, growing in macrophages causing a widespread infection. Some bacteria may set up chronic infections in the bladder. At first the patient has few symptoms; these take about a week to develop but may last for several weeks. Typically the sufferer has a fever and may have diarrhoea and lose blood. The bacteria multiply and are excreted in the patient's faeces and urine for several weeks. Some bacteria leave through blood vessels in the gut wall and peritoneum, which is the membrane lining the abdominal cavity. This damages the blood vessel walls and the patient haemorrhages. Typhoid is diagnosed by detecting the presence of antibodies in the serum and the presence of bacteria in the faeces. The disease is fatal in about 10 per cent of untreated cases.

Treatment and control

The bacterium is sensitive to the antibiotic chloramphenicol but unfortunately strains are evolving which carry a resistance plasmid. A small number of patients become carriers who have no symptoms but harbour the bacteria in their gall bladders, excreting bacteria for months or even years. Many outbreaks have been caused by carriers who were both unaware that they were carriers and unhygienic in their habits, contaminating the food they handled with typhoid bacteria. These people can be treated with a combination of drugs – a sulphonamide with trimethoprin.

The large-scale control of the disease lies in improved sanitation and water treatment to reduce the likelihood of transmission in areas, like that in Fig 8.6, where the disease is endemic and water contaminated. There may be a need to improve the hygiene standards of food handlers. There are vaccines available to protect people, including a whole cell vaccine made from a homogenised suspension of killed *S. typhi* cells. There is also an oral vaccine available which is a mutant strain of bacteria lacking particular enzymes. These bacteria are enclosed within a synthetic capsule which is slowly degraded allowing them to survive and enter the stomach mucosa. There are disadvantages to both of these vaccines, and a new vaccine, free from side effects, has been developed made from fragments of sugar-based polymers found in the bacterium's capsule.

Fig 8.6 Rivers may be the only water supply for drinking, washing and sewage disposal. In these areas water-borne diseases such as cholera, bilharzia and guinea worm are endemic.

QUESTIONS

8.4 What other measures, apart from a programme of vaccination, could be used to limit the spread of typhoid?

8.5 Choose one bacterial disease and draw up a table to link the time sequence of the infection suffered by the patient with the activity of the bacterium.

8.6 The chief medical officer for the town of Greater Hartwell has been notified by GPs in his district of 15 patients who were taken ill with food poisoning. Nine of these patients had attended a dinner at a local hotel, one patient worked as a washer-up in the hotel kitchens, and two other patients were the family of a guest at the dinner. The menu that evening was prawn cocktail with home-made mayonnaise, steak with jacket potatoes and salad and raspberry gateau to follow. What steps do you think the chief medical officer should take to identify the source or sources of infection and to prevent further spread of the disease?

8.5 VIRAL DISEASES

Viral infections cause more concern than bacterial diseases as there are few anti-viral chemicals available. Virus diseases are controlled by vaccination and by blocking transmission. Table 8.3 lists some important viral pathogens. The commonest viral infections are caused by some of the Picornaviruses, the Rhinoviruses, which cause the common cold. There are many different varieties, unfortunately, and immunity is short-lived, so there is little hope of preventing colds in the near future.

Table 8.3 Summary of viral diseases mentioned in the book

Disease	Organism
infectious mononucleosis	Epstein-Barr virus
chickenpox	Herpes varicella
smallpox	Variola
polio	Poliomyelitis virus
pneumonia	Cytomegalo virus
AIDS	Human T-cell leukaemia virus
cold sores	Herpes simplex type I
yellow fever	Togavirus

Another common group are the Herpes viruses. These cause diseases ranging from the fairly innocuous cold sores to viral pneumonia. These viruses are unusual because when they infect some migrate through neurones to the central nervous system where they integrate into the DNA of neurones, making a **latent** infection. Later, perhaps years later, stress or another stimulus causes the virus to emerge and travel down the neurone to cause problems again. People with cold sores for example, caused by Herpes simplex type I, may only need to go out walking on a bright, cold winter's day to reactivate the infection. Similarly some children with chickenpox, caused by Herpes varicella-zoster, can suffer later from shingles which are painful blisters in the skin served by the nerve with the latent virus.

Influenza

Influenza, or 'flu, is a well-known infection which is around all the time but occasionally bursts out in epidemics and pandemics. Healthy people suffer then recover after a week or so but many very young, old or run-down people can die during 'flu epidemics, particularly if complications such as pneumonia set in. Thousands of deaths occur every year, even in the most prosperous countries; the 'flu figures for England and Wales can be seen in Fig 8.7. The disease also causes huge economic losses through so many people having to have time off work because of it. There are three main strains but only type A causes severe pandemics; B and C are less serious. The virus can infect a wide range of other mammals too.

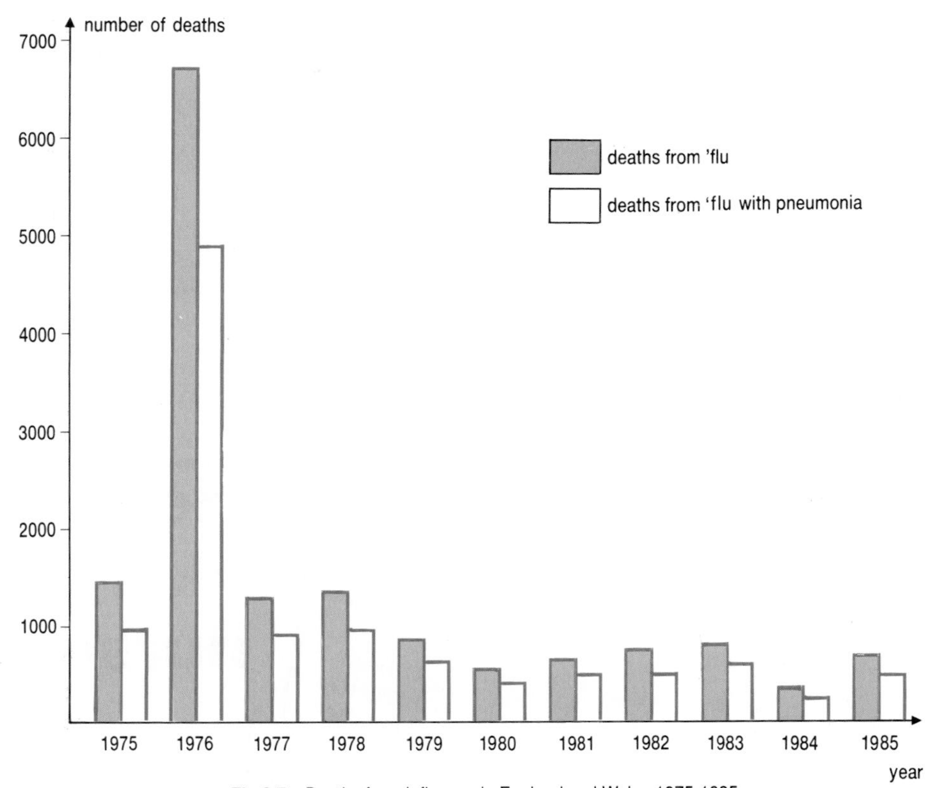

Fig 8.7 Deaths from Influenza in England and Wales 1975-1985.

The disease

'Flu is caused by a myxovirus which is a helical, enveloped RNA virus; unusual because its RNA is in several strands instead of a single strand. 'Flu viruses are shown in Fig 8.8. The envelope is derived from host cell membrane with two projecting glycoproteins which are important in infection. The haemagglutin protein binds the virus to cells lining the respiratory tract; the neuraminidase helps lyse cell membranes allowing the virus entry. These proteins are antigenic and the victim forms antibodies to them. Unfortunately, though, the antigens change frequently so pre-existing antibodies may not be effective against a new infection.

The incubation period is one to three days, then serious damage to the mucous membranes starts. Bacteria such as *Staphylococci* often set up secondary infections in these damaged tissues. Lymphocytes release interferons which protect some cells from infection, causing aches, pains, chills and a runny nose which add to the victim's suffering. Viruses are exhaled, and droplets of moisture containing viruses are sneezed out at a great rate and transmitted to other people.

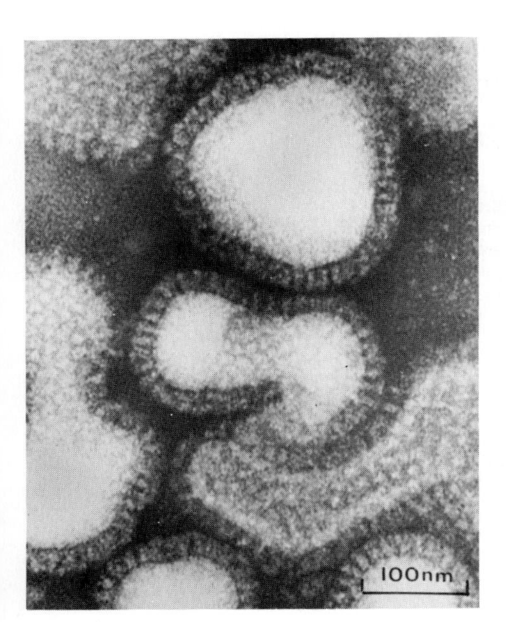

Fig 8.8 Influenza virus particles are not always a regular shape, many may be filamentous. Glycoprotein spikes can be seen projecting through the membrane layer.

Treatment and control

There is little that can be done for the 'flu victim apart from good care, though there are drugs available to treat infections, such as amantadine and rimantadine, if the patient is seriously ill. Taking care of the victim reduces the chances of secondary infections which would need antibiotic treatment.

There is no way of blocking transmission of the virus, apart from wearing masks, and although there are vaccines they have to be redeveloped every few years because of the changing antigens. The haemagglutin antigen changes in five or six-year cycles; the other changes less regularly. New strains can be recombined with a fast-growing cultured strain to produce new vaccine quickly, and vaccines are effective if they are repeated every year with the current strain. This practice protects people most at risk: medical staff, the elderly and those with respiratory problems.

Hepatitis B

Hepatitis B infects huge numbers of people worldwide. Millions die as a result of infection and hundreds of millions remain infectious carriers for most of their lives. In some parts of the world a large proportion of the population is affected, but it is less common in the northern industrialised countries. Where it does occur people are very frightened of catching the infection (see Fig 8.9). It is caused by a virus known as HBV, a spherical virus with a DNA core and an antigenic protein capsid. The virus only infects people and great apes, but its relatives can infect a wide range of other animals.

Hepatitis suit likely after nurse dies

David Brindle
Social Services Correspondent

A HEALTH authority may be sued for negligence towards a nurse ...

Hepatitis threat man jailed

A man who bragged that he had infectious hepatitis as he bit a policeman and spat at others, was jailed for three years at Winchester Crown Court yesterday. It was a false claim but Christopher Knight's conduct was disgraceful and disgusting, said Judge Lewis McCreery, QC.

Knight, 30, of Carmichael Way, Basingstoke, admitted ...

Urgent call for hepatitis vaccine

By Pearce Wright
Science Editor

An epidemic of the hepatitis B virus, which attacks the liver and is a cause of liver cancer, shows signs of easing. The number of acute cases doubled from 1,000 a year to almost 2,000 between 1974 and 1984. But recent figures show a marked fall in the number of cases this year.

Nevertheless, a call ...

Nurses and police demand £40m hepatitis jabs

By David Fletcher, Health Services Correspondent

VACCINATIONS against the blood-borne disease Hepatitis B were demanded by police and nurses on operational duties in separate pleas to the Government yesterday. A vaccine course costs £79 and the cost of treating all nurses and operational police is estimated at about £40 million.

Leaders of both police and nurses have said that their members come into contact with carriers of the disease, which can be fatal. Protection was essential.

Nurses in high risk areas, such as renal units and drug abuse clinics, are given free injections of the vaccine, but not routinely. Some nurses say they have been refused it.

... nurse may feel ill for a while and then recover, without realising that she has become a carrier. Twenty years later she may develop cirrhosis of the liver or cancer," said Mr Goodlad.

Hepatitis tests for immigrants urged

Immigrants from African and Asian countries arriving in Britain should be screened for hepatitis B to help control the spread of the disease, according to a leading expert.

A voluntary screening system should ...

"It would be very reasonable for people coming to Britain from Africa, Asia and the Far East to be tested", Dr Williams said.

"Those who are infected can be treated and their children vaccinated against the disease". He said the Department of Health ...

Gays and prostitutes also seen as most at risk from Hepatitis B virus

Plan to immunise addicts against liver disease

Mothers 'should be tested for Hepatitis B'

Pregnant women should be screened to find out if they are carriers of the liver-damaging disease Hepatitis B, Dame Sheila Sherlock, Professor of Medicine at the Royal Free Hospital, London said yesterday. Bbabies of carriers could then be vaccinated against it.

Dame Sheila said a baby boy whose mother was a carrier had a 50 per cent chance of dying from liver disease. Although the chance was only 14 per cent for girls, they could pass the virus on to their children.

"If you go to Germany every mother is screened for Hepatitis B," she told a press conference in Brussels. It is estimated that 100,000 cases of H... B ...

Hands off alert as hepatitis spreads

Fig 8.9 Hepatitis frightens many people.

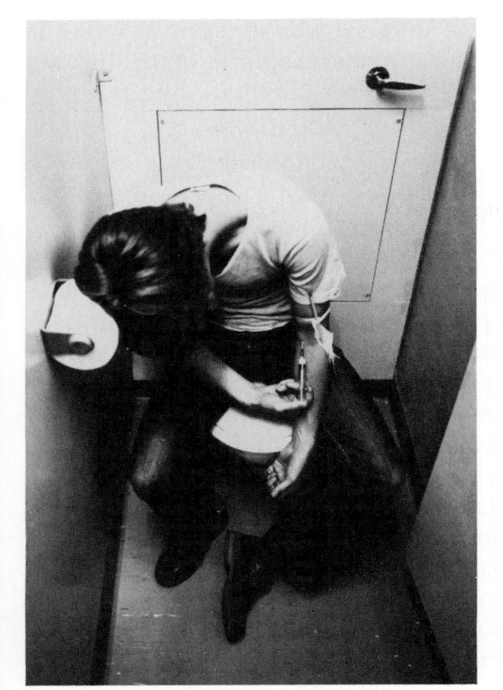

Fig 8.10 In developed countries drug abuse, particularly if it involves sharing needles or unhygienic conditions, is an important means of transmitting Hepatitis B. In developing countries other means predominate.

The disease

The virus is transmitted when infected blood or secretions from one person enter the blood of another person. This can be through cuts and scratches or damaged membranes. One of the commonest ways is by infected mothers passing it to their babies, and many infected babies grow up to be chronic carriers. Other common means of transmission are by sexual or other intimate contact of damaged membranes which lets blood pass from one person to another, and in many countries, by contaminated hypodermic needles, either by drug users sharing needles (see Fig 8.10), or, in poor countries, by improperly sterilised needles or re-used needles in medical treatment. Tattooing, ear-piercing and acupuncture have to be done in the most hygienic way to prevent virus transmission. Some victims were infected by contaminated blood used in transfusions but blood donors are now screened. Medical staff are sometimes victims: nurses and doctors accidentally inoculate themselves when handling contaminated medical equipment.

The symptoms appear a few months after infection and at first are very vague – fever, tiredness, loss of appetite and headaches – but jaundice follows. Bile pigments from the breakdown of haemoglobin accumulate in the skin and liver cells die. Some patients suffer very severe infections which kill them, but others have mild infections and recover after a few weeks, but they shed virus for some time. About 10 per cent of sufferers die and another 10 per cent become carriers with virus in their blood and secretions such as tears. Carriers may suffer more bouts of hepatitis or develop chronic liver failure or liver cancer.

Treatment and control

As there is no real cure for hepatitis controls are focused on preventing infection. People who are in contact with sufferers or carriers regularly are vaccinated to give them long-term protection. There are two vaccines. One is a traditional vaccine made from the blood plasma of chronic carriers. This carries antigen and stimulates antibody production. Human plasma is not cheap so the course of vaccinations may be too expensive for poorer countries to use on many people. Genetic engineering techniques have been used to synthesise the capsule antigen by transferring the genes for it into microbial cells. The new vaccine uses microbially produced antigen; as this can be produced on a large-scale, this vaccine should be cheaper. Short-term protection with antiserum can be given to people who might have been exposed to the virus, for example a doctor who had scratched herself with an infected syringe.

Rabies

Rabies is widely feared, but though there are no accurate figures for deaths each year, they are thought to be only a few thousand worldwide. A few people die each year in many countries; in some places such as India and the Philippines there are more victims. This is very different to the other diseases in this section which cause the deaths of hundreds of thousands or millions each year, but these do not cause us so much concern – why? Perhaps it is the image of the commonest source of infection: a dog, teeth bared and slavering attacking a defenceless person. Or is it that it can take months for the disease to produce symptoms and then it is very difficult to treat?

Once established, rabies is almost certain to result in the death of the victim, though if the disease is treated immediately after infection it can be brought under control. It is one of the few diseases with such a poor outlook. Rabies is no longer endemic in Britain; the two or three cases reported each year are from people bitten by pets which have caught the disease in another country, or people who were bitten whilst abroad.

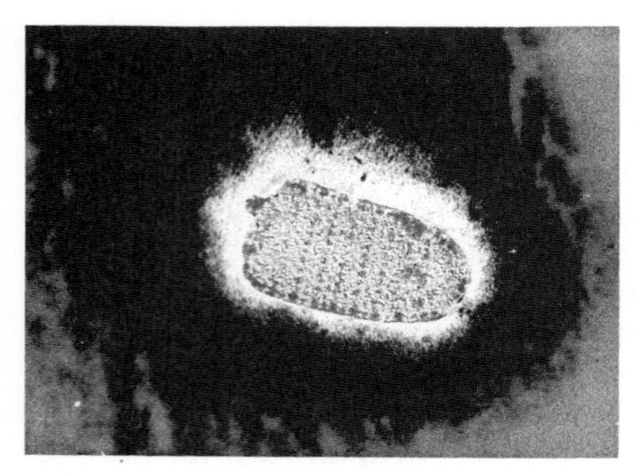

Fig 8.11 The protein coat of the rabies virus can be clearly distinguished.

Though in many countries the local carnivorous wildlife spreads rabies it has not been so in Britain; foxes seem to have been fairly free of the disease.

The disease

Rabies is caused by a bullet-shaped, enveloped RNA virus, shown in Fig 8.11. It is a worldwide zoonosis which can infect all homoiothermic animals but is found most often in bats and dog-like animals such as wolves and jackals. The disease is passed to people most commonly by dogs; one estimate in the USA stated that 15 per cent of children between 2 and 10 years old have been bitten by a dog – a significant proportion of the population. People contract rabies after being bitten or licked by an infected animal and about 50 per cent of these contacts result in infection. The virus is carried in the animal's saliva into teeth marks, cuts and grazes or through a mucous membrane, and the greater the damage the more likely the disease will develop, particularly if the bite is on the face. Though people usually catch rabies from dogs, in rural areas it has been caught from wild animals such as skunks, bats and racoons.

The virus causes damage to nerve cells but does not become active straightaway. Symptoms appear from two or three weeks upto about five months later. The virus migrates from neurone to neurone, killing cells and causing loss of neurone function. Quite quickly the virus can be found within brain cells. The early symptoms are similar to those of 'flu: tiredness, depression and fever, and the patient may well have forgotten about being bitten months before. Once the disease starts, nerve damage, inflammation of the central nervous system and paralysis quickly set in. Patients suffer painful throat and chest spasms when they try to swallow liquids. This is the cause of the hydrophobia – it is too painful for the patient to swallow. Damaged brain tissue causes a variety of other symptoms including personality disorders. Damage to the respiratory centre controlling breathing eventually causes the death of most patients, often within a week. By then virus can be found in most tissues of the body, and in the secretions of tear and salivary glands.

(a)

(b)

Fig 8.12 Some animals suffer furious rabies **(a)** biting and snarling at imaginary things; but others undergo a different sort of character change in dumb rabies **(b)**.

Animal disease

In animals the disease does not kill all its victims – a small proportion survive the infection. Most animal infections follow a similar pattern to human infections, damage to the central nervous system bringing on the strange behaviour, initially distress then more excitable behaviour such as running and biting (see Fig 8.12(a)). Virus is found in the salivary glands at the same time as the animal starts to show signs of brain disturbance. About a quarter of dogs show the behaviour associated with 'mad dogs' – leaving familiar surroundings and travelling long distances, biting indiscriminately and infecting a wide variety of animals. Journeys of over one hundred miles have been recorded. It seems that the extensive brain damage causing this behaviour is necessary to keep the transmission chain going.

Some animals suffer 'dumb' rabies not the 'furious' form, these different patterns of behaviour being determined by the particular strain of virus causing the infection. Pet dogs may become more affectionate, licking their owners and thus passing on the infection, before becoming irritable then suffering a creeping paralysis and eventually collapse. In Europe foxes and cattle are the most frequent victims. Foxes tend to give the infection to each other, whereas cattle are curious about other animals behaving oddly and going up to them and nuzzling them. If the animal they are nuzzling is a rabid fox they get bitten and fall ill very quickly but as cows rarely bite they do not often transfer infections to other animals.

Treatment and control

No-one has survived a full rabies infection, but if bite victims are treated with vaccine, antisera and disinfectants such as cetrimide as soon as they are bitten they have a good chance. If treatment is delayed, even by a few days, the outlook is poor. As the virus has a long incubation period the vaccine can be given immediately after a bite and the body has time to generate an immune response before the virus gets going. The current vaccine is a suspension of killed virus from human cell tissue culture which provokes fewer side effects than the older vaccines containing brain tissue. People who are at risk, such as animal handlers in quarantine kennels and nursing staff, can be protected by a course of vaccine before they are exposed to a potential rabies sufferer.

Where rabies is endemic the disease can be controlled by a vaccination policy. If at least 70 per cent of dogs are vaccinated the transmission links from animal to animal can be broken as there are not enough new victims to keep the disease going. This policy has worked well in the USA. In geographically isolated countries such as Britain the disease can be eradicated by eliminating all infected animals. In rabies-free countries animals entering the country have to be certified as vaccinated before they can be brought in, or may have to undergo a period of quarantine. The quarantine has to be long because of the long incubation period of the rabies virus.

HOW RABIES WAS CLEARED FROM BRITAIN

In the early part of the nineteenth century rabies was found in most parts of Britain with local outbreaks occurring from time to time. These caused anxiety where they happened but were of no great concern elsewhere. Eventually sufficient public concern was raised to take measures to control dogs.

1864 There was a large increase in the number of rabid dogs, particularly in the north. Over 1000 dogs were put down in Liverpool in one month.

1865 The disease spread more widely, and many people were reported as dying of the bite of rabid dogs.

1866 People died in larger numbers in London as well as the north.

1867 Parliament passed the Metropolitan Street Act allowing the seizure of stray dogs in London. The number of rabies deaths in London declined but the Act did not affect other areas.

1869 Deaths continued in northern England and were reported in Scotland too.

1871 Hunt dog packs were severely affected.

1872 The medical journals record cases frequently. Local justices made orders to confine suspect dogs and fines were imposed on owners whose dogs were off a lead in public.

1874 74 people were recorded as dying from rabies.

1887 Rabies Order passed. Local magistrates could order dogs to be muzzled and kept on leads. The order was largely ignored.

1897 The powers of the Metropolitan Act were given to the police nationwide to pick up and destroy any animal thought to be rabid. The import of dogs and cats was controlled and they were subject to quarantine. Dog licences were needed except for working dogs.

1903 Rabies was virtually eliminated.

QUESTIONS

8.7 Why are virus diseases mainly controlled by vaccinations?

8.8 Which people are most likely to benefit from 'flu vaccination?

8.9 Examine Fig 8.7 which shows deaths from 'flu. In which year do you think there was an epidemic? Estimate the number of deaths that year if deaths from pneumonial complications are included.

8.10 What is the role of antiserum when used to treat a virus infection?

8.11 Why is the quarantine period for animals imported into Britain so long?

8.6 FUNGAL DISEASES

Fig 8.13 A culture of *Trichophyton rubrum* which has been grown from skin scrapings of a ringworm sufferer.

Fungal diseases of animals are called mycoses. Few of them are life-threatening. The most common fungal diseases are those affecting the skin such as ringworm, shown in Fig 8.13, and oral and vaginal thrush. Ringworm, caused by a number of fungi, is shared freely by humans and animals. The fungus infects superficial layers of the skin making angry red blisters. A very unusual fungal infection, but one which is becoming more common, is that of extended-wear contact lenses. Fungal spores germinate on the lenses and hyphae extend through them causing problems on the surface of the cornea. It is prevented by improved hygiene and lens care.

There are some severe fungal infections but fortunately these are rare. They are caused by inhaling the spores of free-living fungi. The spores germinate and grow within the lung, and occasionally they may spread within the body, particularly if the patient has a depressed immune system. Farmer's Lung is an allergic response to a minor infection caused by inhaled spores of *Aspergillus* species. Opportunistic infections by fungi are now a major problem for immuno-suppressed patients.

Candidiasis

The disease

Candidiasis is commonly called 'thrush', and is caused by a yeast-like fungus *Candida albicans* and its close relatives. The organism may develop filaments but reproduces by budding. It is found on mucous membranes and is troublesome in children's and elderly people's mouths, in women's vaginas, and in immuno-suppressed patients. Children can develop the condition spontaneously as the organism is a common commensal, or they may pick it up from their mother, or from other children by sharing dummies. The pathogen grows on the inside of the mouth and tongue making white bodies that are attached to the tissue.

Vaginal thrush affects large numbers of women. Normally the secretions in the vagina are pH 4 or 5, controlling the growth of *Candida* which grows best in very mildly acidic conditions. However when a women menstruates, takes the Pill, has diabetes or is pregnant the pH of her vaginal secretions may rise to between 5.8 and 6.8, which is much more favourable for the growth of the yeast. The vagina then itches or has a burning sensation and produces a discharge.

Treatment is by an antifungal compound, often Nystatin, used as a cream or a lozenge to suck. Foaming lozenges, called pessaries, of Nystatin are inserted into the vagina. Many women control their sugar intake or try to regulate the acidity of their vaginas to reduce minor infections, though this would not cure a major infection.

8.7 PROTOZOAL DISEASES

There are not many pathogenic protozoa, but those that do cause infections are very serious problems. Some are blood parasites but they also cause infections which spread throughout the body. Many are zoonoses carried by insects, so vector control is as important as control of the parasite. A few are carried in food or water. Table 8.4 lists some of the important protozoal diseases. The *Plasmodium* species, causing malaria, and the *Leishmania* species which cause a whole range of diseases are responsible for illness in hundreds of millions of people worldwide.

Table 8.4 Protozoal diseases

Disease	Pathogen	Nature
amoebic dysentery	*Entamoeba histolytica*	invades gut mucosa; water-borne
kala azar	*Leishmania donovani*	carried by sandflies; generalised infection – joints, liver, spleen, immune system all affected
trichomoniasis	*Trichomonas vaginalis*	vaginal infection; sexual transfer
Chagas' disease	*Trypanosomas cruzi*	carried by bugs; fever; infection of heart and other organs
sleeping sickness	*Trypanosomas brucei*	carried by tsetse fly; blood infection

Malaria

Malaria is a very widespread disease. About two hundred million people carry the infection at any given time and about two million of these die each year. Nearly 150 countries have conditions suitable for the vector of the disease, a mosquito of the genus *Anopheles*. Malaria has, however, been eradicated from some of these countries and an enormous effort is being put into control or eradication in the remaining countries. Even though the disease is not endemic in Britain it still costs the National Health Service over £1 million a year to treat people who have caught the disease elsewhere, and to protect travellers.

The disease is not one disease but a collection of infections caused by four closely related protozoa of the genus *Plasmodium*. These are *Plasmodium vivax*, *P. falciparum*, which causes the most severe form, *P. malariae* and *P. ovale*. Other species cause infections in other animals including reptiles as well as birds and mammals. Each protozoan species has a specific mosquito host. The plasmodia are very successful parasites as they can survive and evade host immune responses, multiply rapidly and evolve resistance to most chemotherapeutic agents quickly. They are well adapted to life in animals as different as mosquitoes and humans.

The disease

The life and infective cycle of the parasite in humans is shown in Fig 8.14; the parasite undergoes several changes as it emerges in different forms. Malaria sufferers have bouts of severe fever every few days, together with anaemia and liver and spleen damage. The parasite lodged inside host red blood cells uses haemoglobin as a source of amino acids and respires by glycolysis. It has to be able to alter the red cell membrane functions to take in huge amounts of glucose from host blood and generates large amounts of lactic acid and haem which have to be disposed of. When the parasites are liberated into the bloodstream and infect more red cells the patient suffers a bout of fever. The frequency of bouts depends on which species is causing the infection.

There may be damage to kidneys and other organs by the body's immune system which responds to the parasite antigens in them. Blood vessels can be blocked because of the mass of damaged cells and parasites, in turn causing damage to tissues beyond the blockage. This is particularly dangerous if the blood vessels supply the brain. Many patients recover after a few bouts of fever, but anaemia and exhaustion take their toll. If the victim is malnourished or has other infections, the disease can accelerate death from other causes. In some species, some of the parasites in the liver may remain dormant instead of dividing and form a focus of reinfection months later.

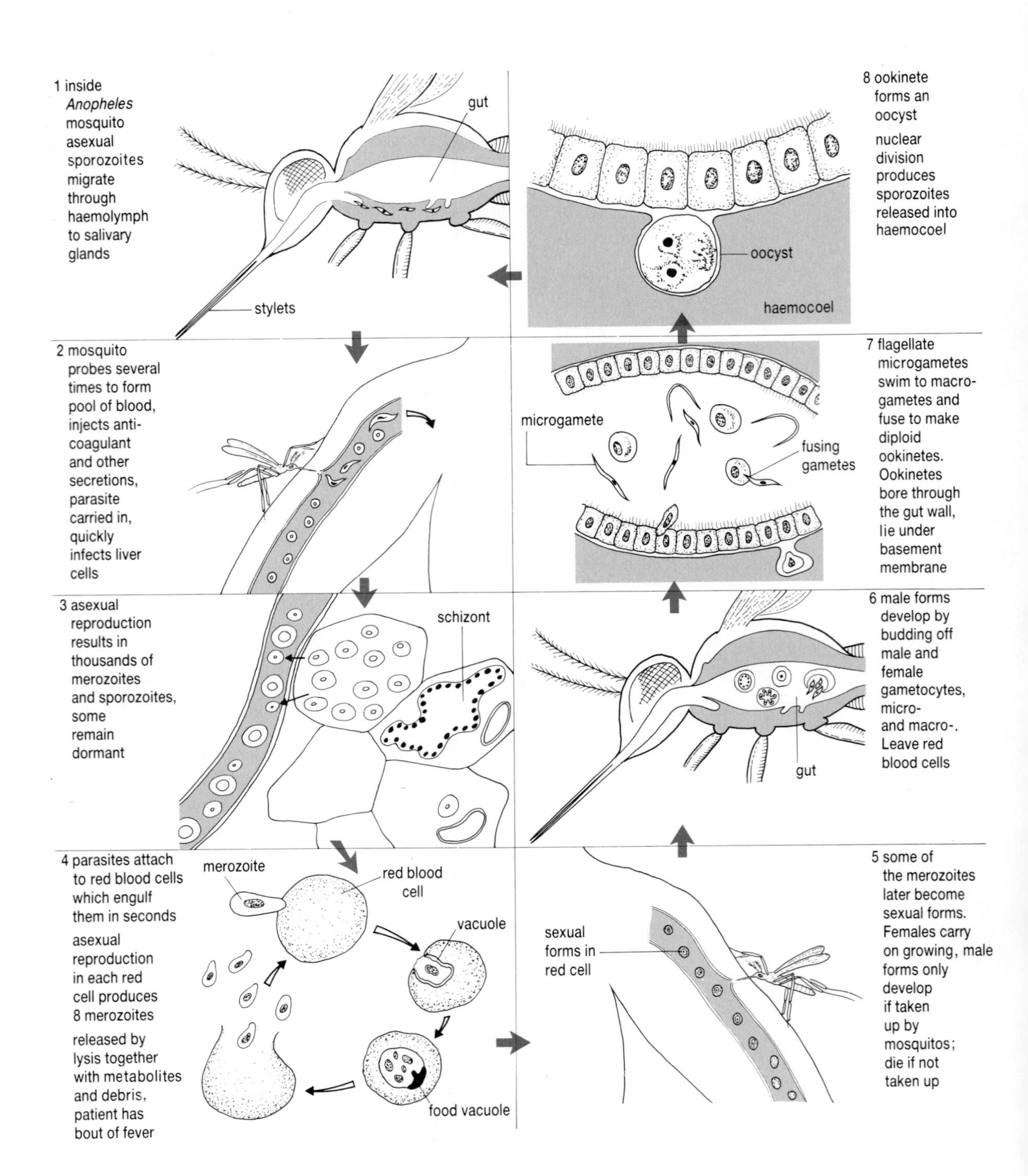

Fig 8.14 Plasmodium life cycle.

Control and treatment

There is no single method which can be used to treat and control malaria. Success depends on a management strategy of prevention of infection with vaccines and drugs, treatment of infected people and control of the vector. After the mosquito has injected the parasites into the bloodstream they are not in circulation for very long, so there is little opportunity for the immune system or drugs to affect them. Also, within liver and red blood cells the parasite is safe from the host immune system. Though antibodies to malaria are formed they do not seem to be very effective at preventing reinfections or relapses; however a patient who has survived several malaria infections does eventually become immune to the disease.

Each stage in the life cycle has different antigens and it is thought that the parasite can change the nature of its surface antigens enough to reduce the effectiveness of pre-existing antibodies. Pregnant women, whose immune system is depressed by their pregnancy, and very young children who are losing the protection given to them by maternal antibody are particulary at risk. Vaccines are not very promising at the moment because of the enormous variation in antigens and the slow response of the immune system. Vaccines containing antigens have been tried but as the antigens in the blood stages are highly specific to each strain an effective vaccine would have to include antigens to many different strains.

Drugs

The main form of treatment is to ensure that a drug, usually chloroquine, is in the bloodstream whenever there are likely to be blood stages present. Drugs can be used prophylactically, killing parasites when they are first injected by the mosquito, or to treat existing infections. Prompt diagnosis and treatment is effective at clearing up the disease. Quinine, together with chloroquine, promaquine and pyrimethamine, have been used to treat malaria and to prevent infections for a long time. Some drugs act against blood stages, others against liver stages, but none are effective against parasites at all stages.

Unfortunately, resistance to the usual drugs is growing. In Thailand, for example, strains of *P. falciparum* were found which were resistant to several drugs. These resistant strains are now appearing in other areas brought in by travellers who are infected in one country then visit another.

The newest treatments are halofatrine and mefloquine, a quinine derivative, which are to be used only on drug-resistant parasites. Mefloquine, the most active drug, is combined with two other drugs, sulphadoxine and pyrimethamine, both of which are active anti-malarial drugs. Malaria parasites can develop resistance to mefloquine so its use is carefully regulated. When combined with the other two drugs, resistant strains do not arise as it is unlikely that an organism can evolve resistance to all three drugs at the same time, but with enough exposure a resistant strain could emerge sooner or later. The use of the triple drug is severely restricted only in order to postpone the inevitable appearance of resistance. The three drugs enhance each other's activity, in other words they are more effective in combination than on their own; this is described as **synergistic** action.

Fig 8.15 Spraying to eradicate mosquitoes is carried out in daytime when the mosquitoes are resting in dark places, such as the eaves of a house. Spraying must be repeated regularly if it is to be successful. Since this photograph was taken workers have been encouraged to use more protective clothing to prevent over-exposure to the insecticide.

Controlling mosquitoes

Insecticides are used to kill mosquitoes. Like the parasite, the mosquitoes have evolved resistance mechanisms and are now resistant to most agricultural insecticides. New insecticides are expensive to develop and buy, and there are environmental problems associated with insecticide residues so the prospects for insecticide control seem poor. Controlling mosquito breeding sites is more promising and has cleared most of Europe of malaria. Mosquitoes breed in open, standing water; their aquatic larvae breathe air from the atmosphere through spiracles fringed with hairs hanging from the surface water tension. If the water is covered then the mosquitoes cannot get in to lay their eggs; if it is moving it is unsuitable for their larvae and if treated with something to reduce surface tensions such as a detergent spray or oil to cover the surface, the larvae will drown.

People can reduce their chances of being infected in an area where malaria is endemic by taking protective measures such as using mosquito netting over beds, insect repellents, screening the doors and windows of their houses, burning mosquito coils and spraying rooms and houses with insecticide at the times the mosquitoes are at rest (see Fig 8.15).

QUESTIONS

8.12 At what stages in a malaria infection is the parasite
(a) susceptible to the host defences
(b) susceptible to drugs in the blood?

8.13 Briefly explain why a vaccination policy is not very effective in controlling malaria.

8.14 Suggest ways in which the vector of the malaria parasite could be controlled.

8.15 Construct a chart summarising the diseases covered in this chapter under the headings 'inductive agent', 'transmission', 'nature of disease' and 'control'.

SUMMARY

Disease can be controlled by treating infected individuals with chemotherapeutic agents, by vaccinating to prevent infection and by environmental management. No single method is effective in isolation; an integrated strategy must be used. Some infections have no easy cure so control rests with prevention.

Diseases of animals are caused principally by bacteria and viruses. Malaria and Hepatitis B, though uncommon in Britain, are amongst the most important diseases globally. Diarrhoeal diseases can cause death by fluid loss. Diseases such as salmonella food poisoning are caused by bacteria found in animals. Vector-carried diseases are controlled by reducing the numbers of vectors as well as curing patients.

Appendix 1

IDEAS FOR PRACTICAL INVESTIGATION

There is scope for a wide range of independent practical or other investigations while studying topics in microbiology. A few suggestions are listed below. Not all of them involve growing or using microorganisms. If you are planning a practical investigation it is vitally important that you discuss your procedure, choice of organisms and growing temperatures with your tutor to avoid potentially hazardous activities. It is generally unsafe to consume any item made in a laboratory, for example yogurt. Some antibiotics produce allergic reactions these should be used as impregnated discs.

Investigation of the activity of acid-producing bacteria

Effect of varying salt concentration on sauerkraut production.
Monitoring acidity changes in sauerkraut production.
Change in pH during the fermentation of milk by *Lactobacillus bulgaricans* to make yogurt.
Monitoring growth curve of *L. bulgaricans* during yogurt making.
Comparison of skimmed, semi-skimmed, sterilised and powdered milk setting qualities in yogurt making.
Effect of adding mineral acid in silage production.
Manufacture of vinegar from beer or wine.
Investigation of oxygen availability on ethanoic acid production.

Yeast

Growth rate and the effect of temperature or pH monitored using haemocytometres, or carbon dioxide production in a dough.
Ability of yeast to ferment a range of sugars and starches.
Investigation of bacterial amylases degrading grain starch for fermentation.
Brewing beer or wine.
Use of bacterial proteases to alter flour quality for biscuit making.
Immobilisation of yeast in alginate in columns and its enzyme activity.

Biodegradation and biotechnology

Decay of biodegradable plastics.
Ability of bacterial enzymes in washing powder to degrade specific materials, effect of incubation temperature, pH, enzyme concentration.
Production of degradative enzymes such as cellulases from soil fungi.
Degradation of oil and water mixtures (*care needed*).
Sequence of colonisation of cellophane buried in soil.
Production of plantlets from callus culture and factors affecting growth rate.
Transfer of genes using *Agrobacterium*.

Fungi

Investigation of fungal growth rates.
Investigation of antibiotic activity.
Making soy sauce or bean curd.
Effect of reducing sugar content in jams to make reduced calorie conserves on keeping properties.
Survey of fungal types in local environs.
Investigation of spore formation in various fungal species.
Growth of edible fungi.
The growth of fungi used in blue cheese making.

Marketing and Information

Investigation of range of microbial products available in local supermarket.
Acceptability of mycoprotein to consumers.
Use of monoclonal antibodies to investigate meat content of 'burgers'.
Effectiveness of various preservatives against spoilage (*care needed*).
Survey of immunisation rates among fellow students and investigation of attitudes.
Newspaper monitoring of salmonella infection outbreaks and the assigned causes.
Planning a campaign to improve the uptake of vaccination.
Production of artwork for campaign materials.
Designing a leaflet to reduce the transmission of a common disease such as hepatitis or salmonella.
Writing a traveller's guide article for a magazine about staying healthy on holiday.
Writing a simple guide to malaria for a businessman visiting Africa or Asia.

Health and disease

Infection of tobacco by tobacco mosaic virus.
Effectiveness of fungicide against damping-off of seedlings in a greenhouse or cold frame.
Sensitivity of monoclonal antibodies in pregnancy testing kits, if HCG available (*expensive*).
Progress of growth of *Monilinia* on apples in warm conditions.
Effectiveness of toothpaste or deodorant as inhibitors of bacterial growth.
Survey of diseases of vegetables sold in shops.

Miscellaneous

Investigation of growth requirements of particular bacteria.
Investigation of the range of protozoa or unicellular algae in local pond.
Comparison of range of species found in two different ponds, identification of abiotic or biotic factors affecting distribution.
Growth of pond photoautotrophs with light of differing wavelengths.
Pigmentation in cyanobacteria with light of differing wavelengths.

Appendix 2

SELECTED ANSWERS TO QUESTIONS

The answers to recall questions can be found by re-reading the relevant section of the text and have not been included, neither have answers to comprehensions, role plays or questions from examination papers.

Investigations: Investigation outlines are not given as there are many ways to carry out an investigation to arrive at an answer. You must check your answer to ensure that you have considered the following points:

1. You have stated exactly what you are attempting to discover and made sure that your chosen procedure will answer that question rather than a peripheral aspect.
2. You have selected suitable apparatus, appropriate to the scale of your investigation. You have stated the time you expect parts of your investigation to take and ensured that you have enough experimental material, numbers of trials or numbers of organisms to give an accurate result.
3. You have ensured that your investigation is a 'fair test'. You must identify the variables involved and include in your account how you propose to control these variables.
4. You must state what you are looking for and exactly what you will measure, for example it is not good enough to say 'I would look to see how well they had grown. I would know that those which had grown the best were in the optimum growth conditions.' It is better to specify measurements; examples are: gain in dry mass, increase in height measured from ground level to shoot tip, and or total root length.

Chapter 1

1.2 kilometre, centimetre, millimetre, micrometre, nanometre. **1.3** A eukaryotic cell would have a visible separate nucleus, if a plant cell there may be visible chloroplasts and vacuole, and stained mitochondria may be seen with a good microscope. A prokaryotic cell would not have any of these features. **1.4** It distinguishes between those bacteria with large amounts of peptidoglycan in their walls and those with little. **1.6** They do not contain cellulose, but contain other compounds not found in flowering plants. **1.11** Green plants absorb light at about 700 nm, at the red end of the spectrum, whereas bacteria absorb more strongly at 850 nm in the far red region. **1.14** No, except to show that cyanobacteria have little peptidoglycan in their cell walls. **1.21** A suspension of conidiospores in sterile distilled water. **1.22** Hydrolysis of the cellulose to release glucose sub-units resulting in the breakdown of fibril structure and therefore strength. **1.25** The fungal hyphae may have penetrated into the wood but are invisible to the unaided eye. **1.33** No, as they do not have the cell wall structure affected by penicillin.

Chapter 2

2.6 Nitrogen sources: (a) sodium nitrate, (b) the air; carbon sources: (a) sucrose, (b) carbon dioxide in the air. The dihydrogen phosphate is a buffer. **2.8** 700 **2.9** Average counts at each dilution: 83.7, 8.3, 0.3. Original numbers for each dilution: 8.37×10^6, 8.3×10^6, 3×10^6. They are not the same as these are random samples from a large population and the 10^{-6} and 10^{-7} dilutions had too few organisms for accurate estimates. The best estimate is that there were $8.4 \times. 10^6$ organisms per cm^3 in the sample. **2.11 (a)** Describe the events in the change from the lag phase to the log phase then describe stationary phase. **(b)** Carbon dioxide or organic acid metabolites such as lactic acid. **(c)** After a short lag phase to induce the enzymes necessary to use the second carbon source the culture would go into a second exponential phase.

Chapter 3

3.4 Very few in (a) and (c), increasingly large numbers in the rest. **3.6** (i) Carbon source (ii) Makes acidic conditions to inhibit other organisms. Lactic acid bacteria. **3.12** The general trend is an increase in the number of cases. The number of cases until 1981 is an underestimate because only those patients who went to their doctor with food poisoning and had it directly diagnosed as such were recorded. **3.15** (i) Dented tins may have seam weld failures which allow micro-organisms to enter the tins from the environment. (ii) While food is defrosted micro-organisms in them grow rapidly; if it is refrozen these become dormant and on further defrosting they can grow to make very high microbial counts. **3.16** It is thought to reduce microbial growth as the food is not held at temperatures favourable for growth for any length of time.

Chapter 4

4.6 (a) All except F. **(b)** Suspended solids (g): A 990, B 399, C 330, D 20, E 168, F 35. A contributes most to suspended solids pollution. A similar calculation shows that C contributes the greatest BOD problem. **(c)** The food factory is likely to be C as the effluent has a high BOD and SS. The cement works is likely to be B as this has a high SS but low BOD. **(d)** Hot water will kill a large proportion of microbial life in a natural watercourse at its entry point. **(e)** Farm D has the higher BOD, indicating a possible leak of slurry.

Chapter 5

5.14 People were worried that the hormone, or a derivative, could be passed on in milk and have harmful effects on the consumer. The trials would be biased if people did know which milk they were consuming. Many people would choose not to buy the treated milk, imposing a financial penalty on farmers using the hormone. Fewer farmers would therefore want to use the hormone in the trial and this would lead to problems associated with small sample size. The sample of people using the milk would not be a random sample, making it far more difficult to monitor any potential effects on consumers. It would be difficult to know how much weight to assign to ill effects suffered by people who knew they were drinking milk from hormone treated cattle, as people may 'generate' symptoms if they are expecting to suffer even if they are not actually ill. **5.28** Use autoradiography to detect the position of radioactive particles in the body.

Chapter 6

6.7 (a) The fungus reduces the translocation of photosynthesised material from infected leaves to shoots and roots. It also promotes the import of materials from other areas of the plant. **(b)** A better control for plant **(c)** would be one where the radioactive carbon dioxide was available to the same leaves in the uninfected plant as in the infected plant. **6.31** Though the overall number of cases and deaths has declined significantly, the likelihood of dying having contracted measles has not declined in the same proportion. The chance of a child suffering nervous system damage is also a cause for concern in measles infections.

Chapter 7

7.18 Storage was at air temperature which could be hot enough to inactivate the virus.

Chapter 8

8.3 To allow comparison between populations of different sizes. Factors leading to the decline of diphtheria include availability of vaccination, improved sanitation and better housing, improved nourishment, availablity of medical care. Not the availability of antibiotics as these did not become widely available until the very end of the time quoted. **8.6** Obtain samples of all the foods eaten at the dinner and analyse the bacteria present, paying attention particularly to the uncooked foods such as prawns, home-made mayonnaise, salad and gateau. Find out what was eaten by the various patients and establish any links between the three patients who were not connected with the dinner and the rest. Establish the procedures used to prepare the meal in the hotel, for example whether salad was prepared in bowls previously used to hold raw chicken but not adequately washed. Establish the suppliers and sources of foods eaten and warn the suppliers of possible hazard. Examine the sources of any foods found to contain food poisoning organisms. If any prove to be sources of the infection improve hygiene standards at the suppliers. Close the hotel kitchens until a thorough disinfection has been carried out. Ensure all hotel kitchen staff and their families are clear of infection and remind of hygienic practice. Try to establish the source of the infection of those not at the dinner. If the fault was at the suppliers establish where other batches of foods may have gone to. **8.9** 1976, 11 500.

FURTHER READING

There are many books available in reference libraries which could widen and deepen your knowledge of micro-organisms and biotechnology. Most are written for college and university students, but can be understood by students studying at A level. Almost every issue of *New Scientist* carries relevant articles and so does *The Biologist*, the journal of the Institute of Biology. For those interested in health studies, *World Health* , the magazine of the World Health Organisation, carries many interesting articles, written for the ordinary reader.

The books listed below would help in a wider reading programme. Some are out of print and will therefore only be found in libraries. The figures in brackets refer to the chapters to which the book is relevant.

General textbooks

Schlegel, *General Microbiology*, 6th edn. Cambridge University Press (densely packed).

Stanier, Ingraham, Wheelis and Painter. *General Microbiology*, 5th edn. Macmillan.

(Much more readable than previous editions.)

Studies in Biology series, Edward Arnold

These are economically-written, single subject books containing far more information than you will need, but invaluable reading.

Baker. No. 138 *The Biology of Parasitic Protozoa*, (1, 8)
Butcher and Ingram. No. 65, *Plant Tissue Culture*, (2, 5, 7)
Deverall, No. 17, *Fungal Parasitism*, (1, 7)
Dixon, No. 44, *The Biology of Aphids*, (7)
Fay, No. 160 *The Blue-Greens*, (1)
Inchley, No. 128 *Immunobiology*, (5, 6)
Noble and Naidoo, No. 111 *Microorganisms and Man*, (3–8)
Postgate, *Nitrogen Fixation*, New Studies in Biology Series (1, 7)
Sharp, No. 82 *An Introduction to Animal Tissue Culture*, (2, 5)
Smith, No. 136 *Biotechnology*, (4, 5)
Warr, No. 162 *Genetic Engineering in Higher Organisms*, (5)

Single topic texts

These may be very comprehensive, containing a wealth of information beyond that needed at A level. Some are academic texts and may be difficult for you to read, so you will need to use your abstracting skills and powers of discrimination. These are just a few of the many good texts available in your local reference library. Books are shelved according to their Classification Number; many are classed as microbiology books (576) but you will also find good texts in sections on plant pathology (581), drugs and therapeutics (615), public and environmental health (614), food technology (664), agriculture (633), biomass energy (662) and waste management (628).

Allsop and Seal, *Introduction to Biodeterioration*, Edward Arnold (3–5)
Atlas and Boulton, *Microbial Ecology – Fundamentals and Applications*, Addison-Wesley (1, 2)
Banwart, *Basic Food Microbiology*, AVI Publishing (2, 3)
Bungay, *Energy – The Biomass Options*, Wiley (4)
Cheremisinoff and Ellerbusch, *Biomass – Applications, Technology and Production*, Dekker (4)

Clegg and Clegg, *Man against Disease*, Heinemann (8)
Cooper, *General Immunology*, Pergamon Press (6)
Diener, *Viroids and Viroid diseases*, Wiley (1, 7)
Edwards, *The Microbial Cell Cycle*, Nelson (1, 2)
Forster, *Biotechnology and Waste Water Treatment*, Cambridge University Press (4)
Fraenkel-Conrat & Kimball, *Virology*, Prentice-Hall (1, 2, 6–8)
Frazier and Westhoff, *Food Microbiology*, McGraw Hill (3)
Gareth Jones, *Plant Pathology, Principles and Practice*, Open University (6, 7)
Lynch and Hobby, *Microorganisms in Action, Concepts and Applications in Microbial*
Ecology, Blackwell (1, 2)
Mantell, Matthews and McKee, *Principles of Plant Biotechnology*, (2, 5, 7)
Mudrack and Kunst, *The Biology of Sewage Treatment and Water Pollution*, Ellis Horwood/Wiley (4)
Richards, *Microbiology of Terrestrial Ecosystems*, Longman (1, 2)
Watson, and Benjamin, *Molecular Biology of the Gene*, (1, 5)
Withers and Alderson, *Plant Tissue Culture and its Agricultural Applications*, Butterworths (2, 7)

Wider reading

These books are not essential reading but they put the work in context and make it easier to understand. Some are written for the general reader rather than scientists.

Baxby, *Jenners Smallpox Vaccine*, Heinneman (8)
Cherfas, *Man made Life*, Blackwell (5)
Communicable Disease Statistics, Office of Population Census & Surveys, Monitor MB2, No. 112 and subsequent editions (Yearly statistics,) (3, 8)
Cross, *Grow your own Energy*, Blackwell/New Scientist (4)
Dixon, *Invisible Allies*, Temple Smith (1–8)
Duclaux, *Pasteur– The History of a Mind*, Library of New York Academy of Medicine/Scarecrow Reprints, (3, 6, 8)
Kaplan *et al. Rabies: The Facts*, Corgi/Oxford University Press (8)
McNeill, *Plagues and People*, Blackwell (8)
On the state of the Public Health, Annual report of Chief Medical Officer. HMSO (4, 8)
Reid, *Microbes and Man*, BBC Publications (8)
Scott, *Pirates of the Cell*, Blackwell (1, 2, 6–8)
Waterson and Wilkinson, *An Introduction to the History of Virology*, Cambridge University Press (1, 2, 6–8)
Webb and Lang, *Food Irradiation – The Facts*, Thorsons (3)

Index